Fundamentals of Environmental Science and Engineering

环境科学工程基础

Shuangying Zheng
郑爽英　主　编

Nguyen Thi Kim Thai
阮金泰[越南]　副主编

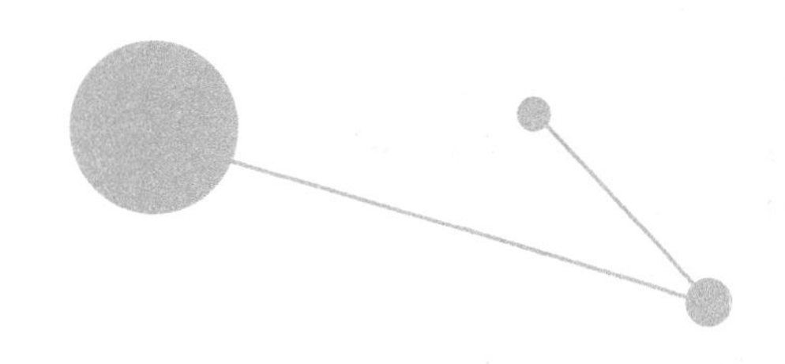

西南交通大学出版社

Science and Technics Publishing House

图书在版编目（CIP）数据

环境科学工程基础 = Fundamentals of environmental science and engineering：英文 / 郑爽英主编. —成都：西南交通大学出版社，2015.7（2021.6 重印）

ISBN 978-7-5643-3824-4

Ⅰ. ①环… Ⅱ. ①郑… Ⅲ. ①环境科学－高等学校－教材－英文 Ⅳ. ①X

中国版本图书馆 CIP 数据核字（2015）第 060170 号

Fundamentals of Environmental Science and Engineering

环境科学工程基础

Shuangying Zheng
郑爽英 主 编

Nguyen Thi Kim Thai
阮金泰[越南] 副主编

责 任 编 辑	赵玉婷
封 面 设 计	墨创文化
出 版 发 行	西南交通大学出版社 （四川省成都市金牛区交大路 146 号）
发行部电话	028-87600564 028-87600533
邮 政 编 码	610031
网 址	http://www.xnjdcbs.com
印 刷	四川森林印务有限责任公司
成 品 尺 寸	185 mm × 260 mm
印 张	15
字 数	487 千
版 次	2015 年 7 月第 1 版
印 次	2021 年 6 月第 2 次
书 号	ISBN 978-7-5643-3824-4
定 价	39.00 元

Preface

With the development of society and the improvement of living standard, the contradiction between people's demand for environmental quality and the unsatisfactory state of the environment becomes more acute. Coming into the 21 century, global environmental issues obviously touch human's subsistence and development. Environment protection and pollution control have become one of the focal problems with which many countries concerned in the world.

Facing the austere realities of environmental and social demand, the environmental education for all-round development should be enhanced in every field of production, construction, management, etc. It is an important way of improving the level of environmental diathesis in the whole society that the course of "Fundamentals of Environmental Science and Engineering" is opened in universities.

In order to improve the abilities of Chinese undergraduate students and Southeast Asian countries' students to read environmental literature in English and obtain relevant information, Southwest Jiaotong University and Vietnam National University jointly compile this English teaching materials named "Fundamentals of Environmental Science and Engineering", which systematically expounds the basic concepts, principles and basic methods of pollution control, and discusses the problems of climate change and global warming. The textbook is divided into seven chapters, including introduction, ecological basis, water pollution control, air pollution control, solid waste disposal, noise pollution, climate change and global warming.

This textbook is used for special foreign language teaching of junior or senior students in colleges and universities whose major is environmental science or environmental engineering. Meanwhile, it can also be used as a bilingual teaching material of non-environmental specialty in colleges and universities for the course of "Fundamentals of Environmental Science or Environmental Engineering", and as a reference book to scientific and technical personnel engaged in environmental protection.

The textbook is chiefly edited by Shuangying Zheng, an associate professor of Southwest Jiaotong University who compiled Chapter One to Chapter Six. While Chapter Seven is finished by Professor Nguyen Thi Kim Thai, the Vietnam National University.

In the compiling course of this textbook, we collected and referred large numbers of domestic and foreign relevant monographs, compilation, teaching materials and other documents. We would like to express our sincere thanks to the authors of the references.

The publication of the textbook is strongly supported by both China Southwest Jiaotong University and Vietnam Science and Technics Publishing House. We own special thanks to Ms. Xue Zhang, the chief editor of Southwest Jiaotong University Press, and Doanh Bui phu, the Director of Educational Administration Division, National University of Civil Engineering, Vietnam. They have started and pushed forward the cooperation between the two publishing houses, and without their help, this book would never have happened. In addition Ms. Yuting Zhao, the duty editor of Southwest Jiaotong University Press, made the textbook be finished as soon as possible. Ms. Yang Liu, the teacher of Southwest Jiaotong University, advanced many valuable opinions about Chapter Six, and the graduate students Hui Ye and Xuetong Liu helped proofread the teaching material. Here, we unfeignedly thank them for their help.

Finally, we wish to acknowledge professor and Dr. Jichun Zhang who gave us the useful help of writing skill.

The book, which will be upgraded in the future, may not be satisfactory to every reader. Limited by time and knowledge of the author, the text, particularly some of the English expressions, might be inaccurate, or may not be rendered in-depth or in detail. For any inaccuracy, please oblige me with your valuable comments and active discussions.

Shuangying Zheng

The way to contact us:

Email: zsy200410@126.com

Tel: (86-28) 8760 0671

CONTENTS

Chapter 1
Introduction

Environmental pollution and ecological destruction are one of the major social problems for mankind to face. In the development process of modern society, many countries pay a high price due to improper handling of environmental problems, resulting in that ecological environment, people's health and social economy suffer great damage. Nowadays, with the rapid development of the global population, industrial and agricultural production and scientific technology, the relationship among population, resources and environment becomes sharp and prominence. The environment problem has increasingly aroused widespread concern and attention, become an important global problem.

But what is environment and environmental problems? In this chapter, some important definitions the main aftereffects of environmental problems, and the research contents of environmental science will be introduced.

1.1 IMPORTANT DEFINITIONS

Environment is the physical and biotic habitat which surrounds us, which we can see, hear, touch, smell, and taste. It includes physical environment and social environment.

Physical environment is the sum of various natural factors surrounding humans. It had existed before human beings appeared, and has experienced a long process of development. The physical environment is composed of non-biological factors and biological factors. The former includes air, water, soil, sunshine, and all kinds of mineral resources etc., without which life-form cannot survive, and the life-form refers to animals, plants and microorganisms. The "environment" in this textbook mainly refers to the layer called biosphere in the physical environment.

Social environment is the result of human production activities over a long period of time. In the long-term development process, people constantly improve the level of material life, science and technology, and cultural life, meanwhile, they create the city and country, industry and traffic, scenic and cultural entertainment and cultural relic, which form the social environment.

Environmental problems refer to the deterioration of the global or regional environmental quality caused by inappropriate human production activities, which is not conducive to the survival and development of mankind.

According to different causes, environmental problems can be divided into two categories, namely, **natural environmental problems** and **man-made environmental problems**. The former, natural environmental problems or the first environmental problems, refers to environmental damage caused by natural disasters, such as volcano eruption, earthquake, typhoon, tsunami, flood, drought and, endemic etc. The latter called man-made environmental problems or the second environmental problems, which refers to the environment pollution, ecological destruction, the rapid increase in population and resources destruction and depletion, caused by inappropriate human production activities, and it will be mainly discussed in the textbook.

System, according to Webster's dictionary, is defined as "a set or arrangement of things so related or connected as to form a unit or organic whole, such as a solar system, an irrigation system, a supply system, the world or universe".

Pollution can be defined as an undesirable change in the physical, chemical, or biological characteristics of the air, water, or land that can harmfully affect the health, survival, or activities of humans or other living organisms.

When the goal of improving environmental quality is taken to be improving human well-being, the word "environment" broadens to include all kinds of social, economic, and cultural aspects. Such broadness is unworkable in many real situations and impractical in a textbook designed for a one-semester course. Our examination of environmental problems is therefore limited by our definition of "environment".

1.2 DEVELOPMENT OF ENVIRONMENTAL PROBLEMS

Environmental problems occur and develop with the emergence and development of human beings. In an agrarian society, people lived essentially in harmony with nature, raising food, gathering firewood, and making clothing and tools from the land. The wastes from animals and humans were returned to the soil as fertilizer. Few, if any, problems of water, land, or air pollution occurred (Figure 1.1). For the small settlements which grew up, the supply of food, water, and other essentials and the disposal of wastes had to be kept in balance with the changing community, but no serious environmental problems were created.

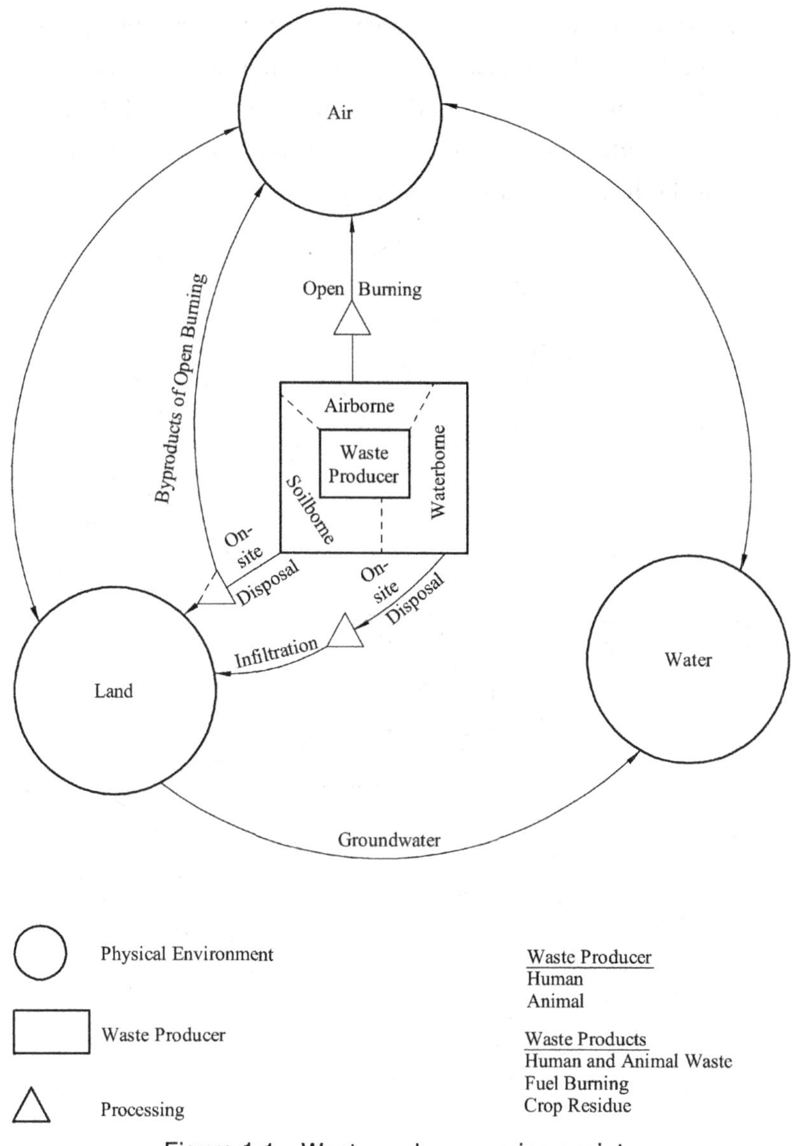

Figure 1.1 Waste cycle – agrarian society

(Source: J. Glynn Henry and Gary W. Heinke, *Environmental Science and Engineering*, Prentice Hall, 1989, P5)

The cities of ancient times had systems to supply water and to dispose of wastes. The municipal technology of ancient cities seems to have been forgotten for many centuries by those who built cities throughout the world. Water supply and waste disposal were neglected, resulting in many outbreaks of dysentery, cholera, typhoid, and other waterborne diseases. Until the middle of the nineteenth century, it was not realized that improper waste disposal polluted water supplies with disease-carrying organisms. The industrial revolution in nineteenth-century Europe and North America aggravated the environmental problems since it brought increased urbanization with the industrialization. Both phenomena, urbanization and industrialization, were and are fundamental causes of water and air pollution which the cities of that time were unable to handle.

Rapid advances in technology for the treatment of water and the partial treatment of wastewater took place in the developed countries over the next few decades. This led to a dramatic decrease in the incidence of waterborne diseases. Figure 1.2 illustrates the waste disposal cycle for an industrialized society. Note that all wastes discharge into the environment, and thus pollute water, air, and land systems.

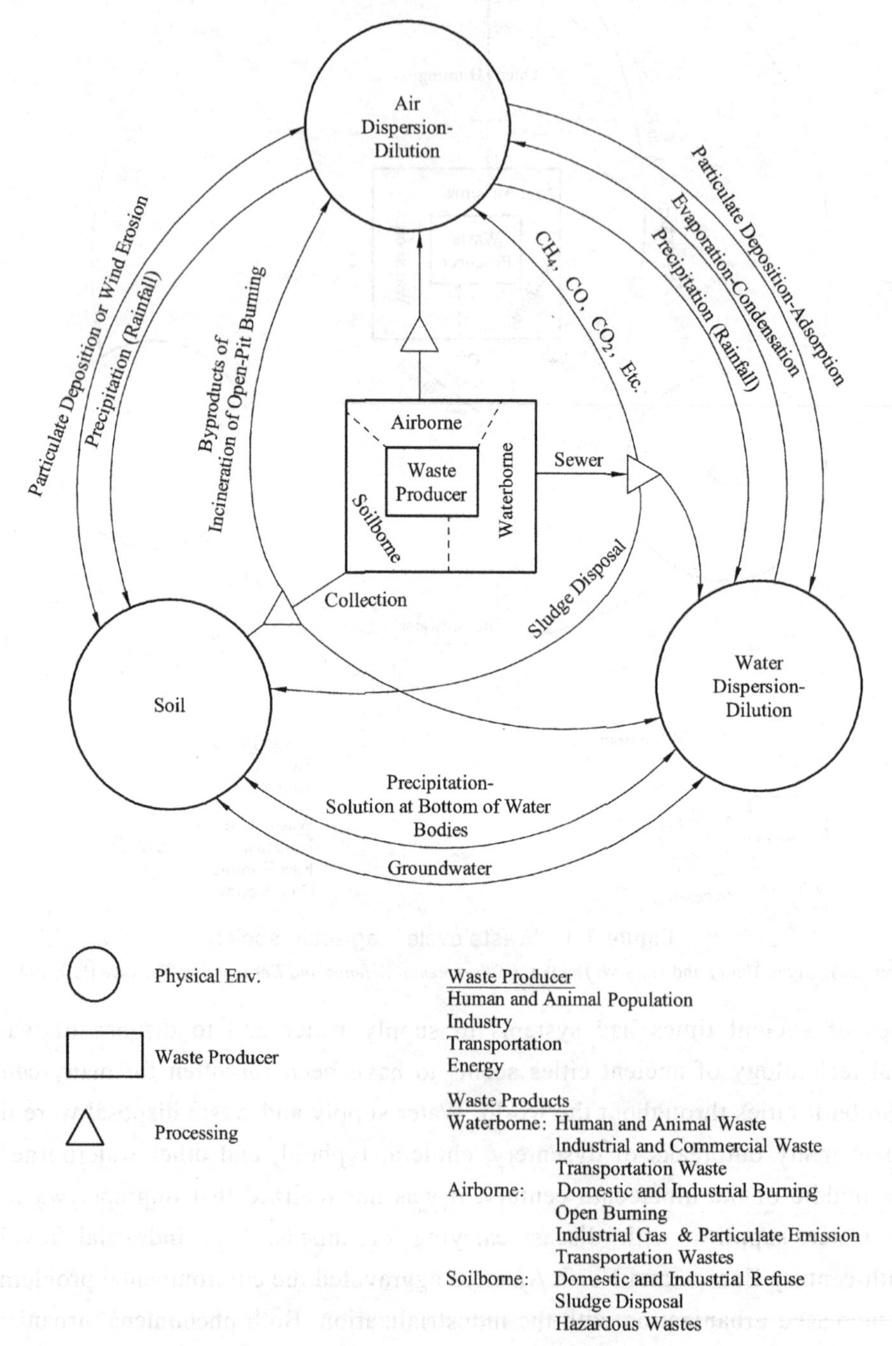

Figure 1.2 Waste cycle – industrialized society

(Source: J. Glynn Henry and Gary W. Heinke, *Environmental Science and Engineering*, Prentice Hall, 1989, P6)

Following the Second World War the industrialized countries experienced an economic boom fueled by a burgeoning population, advanced technology, and a rapid rise in energy consumption. During the 1950s and 1960s this activity significantly increased the quantity of wastes discharged to the environment. New chemicals, including insecticides and pesticides, used without sufficient testing for their environmental health effects, caused, and continue to cause, enormous problems not anticipated when they were introduced. Unfortunately, the problem is worsening as the variety and amounts of pollutants discharged to the environment increase inexorably while the capacity of our air, water, and land systems to assimilate wastes is limited. For example, around 1950's, eight famous incidents of pollution (See Table 1.1) in Belgium, America, Britain, and Japan, have shocked the whole world.

In the recent forty years, with the rapid growth of population and the unprecedented expansion of human impact on earth, there are a series of sharp contradictions among the population, resources, environment, and economic development, which has aroused attention of countries all over the world. In 1972, the first Conference on the Human Environment was held in Stockholm, which was a wake-up call of environmental problems and forced each government to make protection statutes of resource and environment in their national government's schedule. However, although people have made various efforts, the problem has not been really solved. What is more, many new environmental problems, such as the exhausting ozone layer, global spreading acid rain, the biological species reduction, population growth, and lack of resources, are threatening the survival and development of human beings. Serious global environmental realities force people to reflect on the high consumption of resources and the neglect of environmental problems seriously, and try to explore an effective boom way. Under such a background, in June, 1992, the United Nations Conference on Environment and Development (UNCED) was held in Rio de Janeiro, on which a commonly accepted human development road – the road of sustainable development was put forward.

1.3 SUSTAINABLE DEVELOPMENT

1.3.1 Relationship between Environment and Development

In the 1960s, after suffering from a series of serious environmental public hazard incidents, humans began to actively review and sum up the drawbacks of traditional economic development patterns. They found that pure economic growth did not equal to development, and for development itself, besides demanding the growth of "quantity", the more important thing was to raise and improve the overall "quality". In order to obtain freedom in nature, humans must use knowledge to build a better environment under the condition to cooperate with nature.

Table 1.1 The famous eight incidents of pollution

Name	Pollutant	Place	Time	States of poisoning	Symptoms of poisoning	Causes of injury	Causes of the incident
Maas Valley smog incident		Belgian Maas Valley	1930.12	Thousands of people got ill, 60 people died	cough, tachypnea, lachrymation, sore throat, nausea, vomiting, chest stuffiness	People inhaled SO_3 deep into the lungs	(1) Lots of heavy factories in the valley (2) The inversion weather (3) The accumulation of industrial pollutants (4) Foggy days
Donora smog incident	Soot, SO_2	Donora town, Pennsylvania, America	1948.10	About 42% of the people were ill, 17 people died in 4 days	cough, sore throat, chest stuffiness, vomiting, diarrhea	People inhaled sulfate into the lung	(1) Lots of factories (2) foggy days (3) The inversion weather
London smog incident		London, England	1952.12	4,000 people died in 5 days	chest stuffiness, cough, sore throat, vomiting	People inhaled soot adhering to the sulfuric acid foam into the lungs	(1) The residents used high sulfur coal for heating, which discharge lots of dust with sulfur dioxide (2) The inversion weather
Los-Angeles photochemical smog incident	Photochemical smog	Los-Angeles, America	1943.5-10	The majority of the local residents got ill, 400 people who were over 65 years old died	Ophthalmocace, laryngitis	Photochemical smog created by oil industrial and automobile exhaust gas under altraviolet	(1) More than 1,000 tons of hydrocarbon were discharged into the air every day (2) The city was surrounded by mountains in three sides (3) The air flowed slowly in horizontal direction
Minamata incident	Methyl mercury	Minamata town, Japan	1953	More than 180 people were ill, more than 50 people died	lisp, gait instability, facial dementia, deafness, blindness, whole body numb, nervous breakdown	People got sick or even died through eating the poisoned fish	The waste water and residues contained toxic methyl mercury polluted the source of water.
Toyama incident	Cadmium	Toyama, Japan	1931-1972	More than 280 people were ill, 34 people died	Arthralgia, Neuralgia, bone pain, bone softening atrophy, natural fracture	The rice and drinking water contained cadmium	Cadmium-tainted water was discharged into the river by factories without purification
Yokkaichi incident	SO_2, Coal dust, Heavy metals dust	Yokkaichi, Japan	1955-1972	More than 500 people were ill, 36 died of Asthma	Bronchitis, Bronchial Asthma and Emphysema	People inhaled the particles contained toxic heavy metal and SO_2 into their lungs	Factories discharged large quantities of SO_2, coal dust and heavy metal dust into the air
Rice bran oil incident	PCB	23 prefectures, including Aichi Prefecture, Kyushu, Japan	1968	More than 5,000 people got ill, 16 people died	Swollen eyelids, sweating frequently, body covered with red bumps, nausea, vomiting, liver function decline, courbature, badly cough or even death	Eating the rice bran oil contained PCB	In the production process of rice bran oil, the PCB, as a heating medium, was got into the oil because of poor mana- gement

The environment is where we all live, and development is what we all do in attempting to improve our lot within that abode. Mankind is a part of nature and life depends on the uninterrupted functioning of natural systems. Environment and human activities are inseparable. Conservation of nature cannot be achieved without development to alleviate poverty and misery of hundreds of million people. Unless the fertility and productivity of the planet are safeguarded, the human future is at risk.

1.3.2 Definition of Sustainable Development

The concept of sustainable development has been evolving. In "Our Common Future", it was defined as the development that meets the needs of the present without compromising ability of future generations to meet their own needs (WCED, 1987). Broadly speaking, sustainable development strategy is aiming at promoting the harmony between human beings and nature; from a narrow view, sustainable development implies the sustainability of nature, i.e. the inter-generational fair distribution in resource supply and its costs and benefits, including the intra-generational fairness between regions.

The definition has remained a classic one, and this principle was universally accepted by more than 100 countries, including China, at the 1992 UNCED. The World Summit on Sustainable Development (WSSD) in 2002 made it clearer that the economic development, social development, and environmental protection be three pillars of sustainable development at local, national, regional, and global levels, which are interdependent and mutually reinforcing. In another word, it means that economy, society, and the protection of environment should be developed in a balanced way with an aim of developing economy as well as protecting the natural resources and environment, including atmosphere, fresh water, oceans, land and forests, etc. The core of sustainable development is "development", in particular, economic and social development, which could be gained under the precondition of strict control on population, improvement of population quality, environmental protection, and sustainable utilization of resources.

Sustainable development stresses fairness, equity, and integration of the above-mentioned three pillars. It took poverty eradication, changing consumption and production patterns, and protecting and managing the natural resource base for economic and social development as the overarching objectives, and essential requirements for sustainable development.

1.3.3 Measures of Sustainable Development

For different regions and different countries, even in different periods of the same country, the sustainable development measures are different. For most developing countries, economic development and meeting people's basic living needs are the premise of sustainable development. For developed countries, the emphasis should be put in the technique reform, low input, and low consumption.

Human beings must make profound changes in order to realize sustainable development. These changes will come from all areas, such as society, economy, scientific and technological education, administration, and law etc.

First, we should keep population at a sustainable development level. Population is both the most important development resource and the heaviest burden for development. For example, in China, the absolute amounts of many resources rank first in the world and the total size of the economy is also in the forefront, but because of large population, per capita possession of resources and per capita GDP lag far behind those in many other countries. Limiting family size to two children and giving support to international organizations that promote family planning in less developed nations are important ways of controlling population growth. In addition, a national education to promote the citizens' basic scientific literacy is very important. The way and degree of participation of the masses will decide the realization process of sustainable development.

Second, we should conserve natural resources which are the material basis for human survival and development. Natural resources can be categorized as non-renewable resources, such as mineral, petroleum and natural gas etc., and renewable resources, such as forest, grassland etc. To save resources, developing countries must change the traditional mode of economic growth, i.e., the extensive development mode of high input and high consumption changes to the intensive pattern of saving resources and reducing consumption. Meanwhile, developed countries need to solve the problem of high consumption. Western developed countries, whose populations account for 26 percent of the world, consume over 70 percent of the world's merchandise and labor force. But, for the developing countries such as China, it is often a case of too many people sharing too few resources. Table 1.2 largely reflects the huge gap between the developed countries and the developing countries in the aspect of consumption.

Table 1.2 The distribution of the world's consumption

Merchandise		Per capita consumption unit	Developed countries (26% of the world population)		Developing countries (74% of the world population)	
			Proportion of world consumption (%)	Per capita consumption	Proportion of world consumption (%)	Per capita consumption
Food supplies	Energy	calories/day	34	3395	66	2389
	Protein	g/day	38	99	62	58
	Fat	g/day	53	127	47	40
Paper		Kg/year	85	123	15	8
Steel		Kg/year	79	455	21	43
Other metals		Kg/year	86	26	14	2
Commercial Energy		Tonne coal equivalent / year	80	5.8	20	0.5

(Source: "Our Common Future", *World Commission on Environment and Development*, 1987.)

The Conservation of natural resources also means using resources more efficiently. Currently we waste about 50% of the consumed energy. Every day, huge amounts of water and other resources are wasted. Although this wastefulness is often viewed as one of society's greatest faults, it is also one of our greatest opportunities for improvement. By becoming more efficient we can cut waste, reduce environmental damage, and ensure a steady supply of resources for future generations.

Third, we must prevent industrial pollution so as to protect the environment. Industrial pollution is a major cause of environmental pollution and a major obstacle to economic and social sustainable development. In order to reduce and prevent industrial pollution, we must vigorously promote clean production technology, or clean technology in short. Clean technology is a kind of process and technology which can reduce the consumption of raw materials and energy, and effectively prevent the production of pollutants and other wastes. It requires the input of energy and resource to a minimum and the output of waste and pollutant to the lowest degree in the production process. At present, clean production technology is advanced technology in controlling the industrial pollution, and will become the dominant technology in industrial production in twenty-first century.

In addition, human beings also need to take other measures, such as carrying out ecological agriculture, planting more trees and grass, protecting biological diversity, implementing comprehensive renovation on cities' environmental pollution, improving the environmental legal system, promoting the environmental science research and education, strengthening environmental management, international exchanges, and cooperation.

1.3.4 Making Sustainable Development Work

"Developing countries are littered with the rusting good intentions of projects that did not achieve social or economic success," writes Walter V. C. Reid, a leading authority on Third World development. To making matters worse, he notes, ill-conceived projects have wrought considerable damage to the Third World. Erosion, desertified landscapes, pesticide poisoning, pollution, and deforestation are some of the results of a multibillion dollar annual budget for Third World development. Why?

There are many reasons. The first is the good intentions of the international lending agencies, the multilateral development banks (MDBs), and development agencies. There are four MDBs: the World Bank, the Inter-American Development Bank, the Asian Development Bank, and the African Development Bank. They are funded by groups of nations. The World Bank, headquartered in New York City, for example, is mainly supported by the United States, France, Germany, the United Kingdom, etc. MDBs lend billions of dollars a year to Third World nations to finance economic development projects, from road construction to farming to dam building.

The MDBs are joined by private commercial banks and international development agencies. The US Agency for International Development, for example, is a key player. It provides outright grants for development projects.

All this is well and good. Problems arise, in part, because of the types of projects the MDBs fund or support. Many projects are unsustainable. Thus, the MDBs have not only failed to stop environmental deterioration, but they have, in many cases, worsened it. Large dam projects, for example, flood productive farmland, displace people, increase the prevalence of waterborne disease, and reduce sediment that nourishes estuarine life. The costs of these projects often exceed the economic benefits gained from hydroelectric power and irrigation water. Irrigation water in arid regions leads to salinization and waterlogging, both of which can decrease productivity and render land useless. Over half of the Third World's irrigated cropland, for instance, suffers from salinization.

MDBs have helped finance deforestation and colonization projects as well. In western Brazil, forests were cut to make way for farmers and ranchers, but 80% of the people soon left because the soils were quickly washed away or lost their fertility.

Pesticide use to support large agricultural projects funded by MDBs has also proved economically and environmentally costly. In Central America, in fact, chronic and acute pesticide poisoning is among the region's most serious problems.

Volumes of horror stories could be written about well-intended projects gone awry. Fortunately, the MDBs and others are beginning to see the need for new practices for sustainable development.

Considerable gains can be made in the Third World by improving energy efficiency. Simple changes in wood cookstoves, for instance, could cut energy demand drastically and help reduce deforestation in the Third World. Energy efficiency, combined with projects to replant trees and manage forests better, could go a long way toward helping Third World nations meet their needs sustainably. Today, however, energy efficiency projects constitute less than 1% of the international aid.

Agriculture, forestry, and animal husbandry can be made sustainable, but important changes are needed. First, funding agencies must rely more on local knowledge than agricultural experts from the West who often attempt to transfer costly technology to Third World nations. Second, MDBs and other agencies agencies must begin to study the environmental, social, and political effects of proposed projects. Until very recently, the environment and the long-term sustainability of projects have received little or no attention at all. A more careful analysis will help MDBs and others avoid projects that are doomed to fail or destined to create widespread environmental damage. Third, developers must design with nature, rather than continue to redesign nature. "Modification of the environment to fit the needs of a production system," writes Walter Reid, "is much less likely to be sustainable modifications of the production system to fit the environment." That's because most environmental modifications have serious repercussions. Fourth, development should preserve genetic diversity whenever possible. Extractive reserves in Brazil and other countries are a case in point. They can provide sustainable income at a far lower cost than traditional development practices like clearing and cutting forests. Fifth, inappropriate laws and policies must be removed. Subsidies for pesticide use and for ranching in cleared tropical rain forests are two examples of ruinous policies. Sixth, widespread participation among locals should be encouraged. Planners have long ignored the input of people who will be affected by their projects. They have also ignored knowledgeable local experts, who better understand the people's needs, cultures, beliefs, and the environmental constraints of areas being affected. "Planners should not assume that they know people's needs," writes Walter Reid. Development is unlikely to be sustained unless the needs of people are identified and local residents support the project. Seventh, MDBs and other agencies must be more flexible, allowing projects to shift as problems arise. Inflexible bureaucracies often impair projects by refusing to allow for changes, even in the face of serious problems. By funneling money through smaller, non-governmental organizations, the large bureaucracies can be separated from the management of projects, allowing greater flexibility and ensuring greater success.

These changes can help the Third World develop along a sustainable path, operating within the limits posed by the earth. They can also be applied to the developed countries in future development.

1.4 WHAT IS ENVIRONMENTAL SCIENCE?

What is environmental science? To understand the term, we need take each word separately. The word "environmental" refers broadly to everything around us: the air, the water, and the land as well as the plants, animals, and micro organisms that inhabit them. Science, of course, refers to a body of knowledge about the world and all its parts. It is also a method for finding new information. Science seeks exactness through measurement, insight through close observation, and foresight through its theories.

Environmental science comes into existence as a recognized discipline to cope with the vast problems spawned by overpopulation, resource depletion, and pollution. It has become a key tool for our survival.

Modern environmental science is aimed at helping us control our own actions in the natural world to avoid irreparable damage. In this sense, environmental science means learning to master ourselves.

To solve the highly complex problems of overpopulation, resource depletion, and pollution requires a knowledge of many scientific fields. Environmental science calls on chemistry, biology, geology, and a great many other disciplines, including sociology, climatology, anthropology, forestry, and agriculture. Spanning this wide range of knowledge, environmental science offers an integrated view of the world and our part in it. Environmental science takes on the colossal task of understanding complex issues. It is an often awkward melding of science, engineering, and liberal arts that requires broadly educated men and women in the age that leans heavily toward specialization.

Environmental science differs from the traditional "pure", or "objective", science, which seeks knowledge for its own sake. Instead, it offers a great deal of urgent advice and reaches many conclusions that challenge cherished beliefs and practices. You may find this true as you read this book. In contrast to the astronomer in a mountaintop observatory or the cell biologist in a laboratory, environmental scientists are often in the thick of things, and at the heart of today's hottest debates.

Environmental science is the study of the environment, its living and nonliving components, and the interactions of these components. By choice it focuses on the ways that humans affect the environment and the ways our actions come back to haunt us. Crossing many traditional boundaries, it attempts to find answers to complex, interrelated problems of population, resources, and pollution, problems that threaten the welfare and long-term survival of humanity.

Changing our ways will be a colossal task, a process that will take generations to complete. It will involve arduous work in many fields. The moon landing is a weekend home-improvement chore compared with the job ahead. The study of environmental science is a cornerstone of change.

1.5 WHAT IS ENVIRONMENTAL ENGINEERING?

Environmental engineering is an important branch not only of environmental science but also of engineering. The Environmental Engineering Division of the American Society of Civil Engineers (ASCE) has published the following statement of purpose:

Environmental engineering is manifest by sound engineering thought and practice in the solution of problems of environmental sanitation, notably in the provision of safe, palatable, and ample public water supplies; the proper disposal of or recycle of wastewater and solid wastes; the adequate drainage of urban and rural areas for proper sanitation; and the control of water, soil, and atmospheric pollution, and the social and environmental impact of these solutions. Furthermore, it is concerned with engineering problems in the field of public health, such as control of arthropod-borne diseases, the elimination of industrial health hazards, and the provision of adequate sanitation in urban, rural, and recreational areas, and the effect of technological advances on the environment.

Thus, we may consider what environmental engineering does not. It does not concerned primarily with heating, ventilating, or air conditioning (HVAC), nor is it concerned primarily with landscape architecture. Neither should it be confused with the architectural and structural engineering functions associated with built environments, such as homes, offices, and other work places.

There are mainly two aspects in res environmental engineering, from is to protect the environment from being damaged by the human unadvisable activities, and the other is to protect humans from being injured by the adverse environment so as to enable people to live healthily and cosily.

Chapter 2
Ecological Basis

Ecology is the theoretic basis of environmental science. Organismal survival, growth, activities, and reproduction require space which provides material and energy to maintain life. Organismal life space is the biological environment.

With the rapid development of the world economy, the consumption of natural resources and the emission of pollutants are dramatically increased, which aggravates environmental pollution and ecological destruction, and endangers the development of organism and human beings.

But what is the ecological destruction and ecological imbalance? This chapter will focus on some basic concepts of ecology, the structure and function of ecosystem and the effects of ecological law on environmental protection.

2.1 INTRODUCTION TO ECOSYSTEM

2.1.1 Definition of Ecology

The term "ecology" is derived from the Greek "oikos", meaning "house", and combined with "logy", meaning "the study of ". Thus literally, ecology is the study of the earth's household.

For our use, ***ecology*** can be defined as the study of the relationship between organisms and their environment. Here, "environment" is taken to mean both the physical and chemical environment of air, soil, and water, and also the biological environment itself. It takes the entire living world as its domain in an attempt to understand all organism-environment interactions. Given the vast number of organisms in the world, the realm of ecology is immense.

There have been many changes and development about ecology since 1870's. At present, the research of ecology is generally divided into individual ecology, population ecology, community ecology, and ecosystem ecology. This textbook focuses on ecosystem ecology.

2.1.2 The Biosphere

The part of the earth that supports life is called the ***biosphere***, or ***ecosphere***. As is shown in Figure 2.1, the biosphere extends from the floor of the ocean, approximately 11,000 meters (36,000 feet) below the surface, to the top of the highest mountain about 9,000 meters (30,000 feet) above sea level. If the earth were the size of an apple, the biosphere would be a thin layer only as thick as its skin.

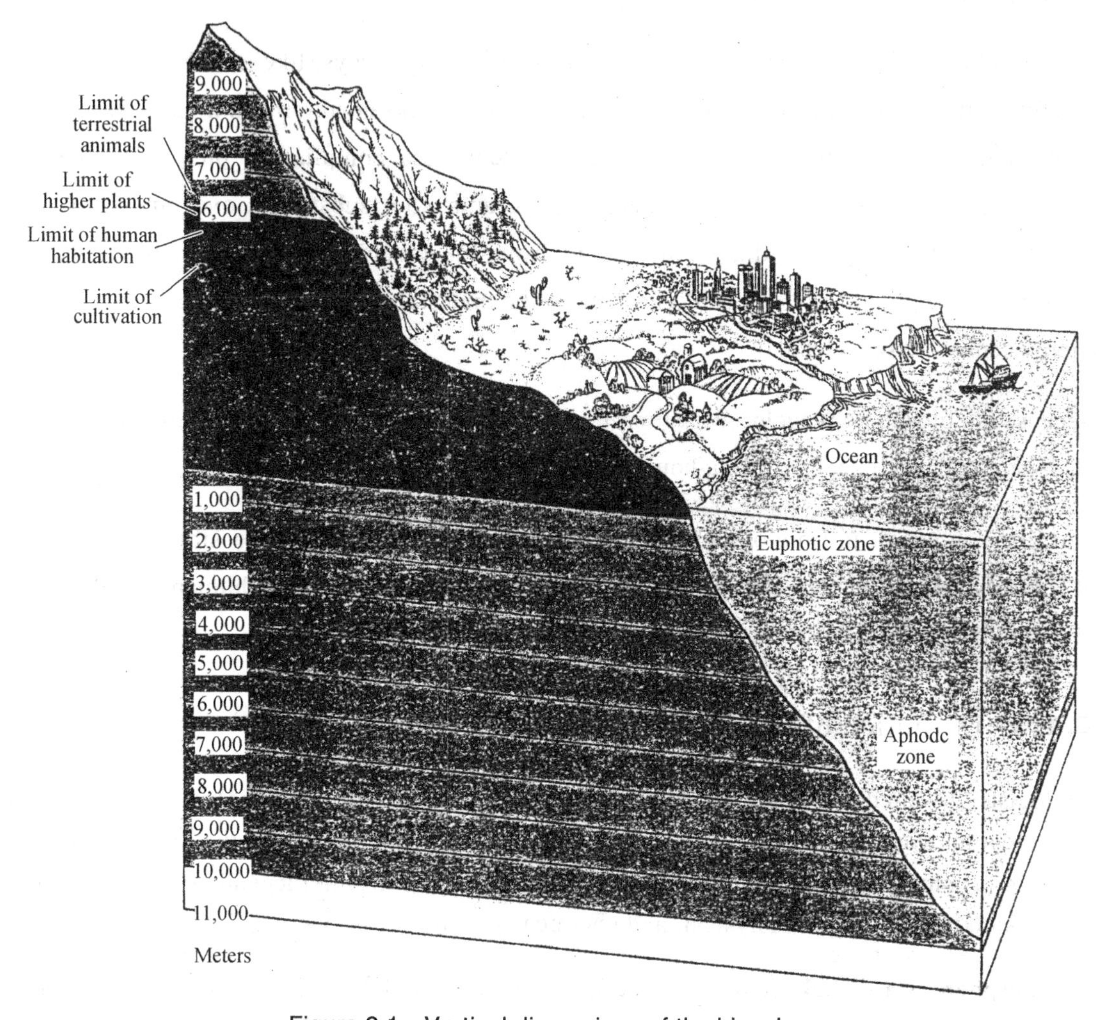

Figure 2.1 Vertical dimensions of the biosphere

(Source: Daniel D. Chiras, *Environmental Science: Action for a Sustainable Future*, The Benjamin/Cummings, 1991, P46)

Life forms are scarce at the far extremes of the biosphere. Life evolved under more moderate conditions than the biospheric extremes, and it is in these conditions that most species thrive. The zone of abundance is a narrow band extending from less than 200 meters below the ocean surface to about 6,000 meters above sea level.

Life exists mostly at the intersection of land (lithosphere), air (atmosphere), and water (hydrosphere). And from these vast domains come the ingredients that make life possible – minerals from the soil, oxygen and carbon dioxide from the air, and water from oceans and lakes. If you put these ingredients in a test tube, you have simply a mixture of air, water, and soil. But when these molecules are uniquely organized in living things, they produce a fascinating array of shapes and forms that, among other things, grow, reproduce, and respond to various stimuli.

The biosphere – this living skin of planet earth – is, in many ways, like a sealed terrarium. If carefully set up with soil, water, plants, and a snail or two, a sealed terrarium operates without interference. The terrarium is a sustainable system, less complicated but more like the biosphere, which recycles water, oxygen, carbon, and other substances over and over in a perpetual cycle. Only one thing must come from the outside for life to continue in the terrarium and in the biosphere: sunlight. Unlike the materials that make life possible, sunlight energy cannot be recycled.

Because the biosphere recycles all matter, it is called a closed system. The closed nature of the biosphere makes all species throughout all time cousins of sorts.

The first law we glean from our study of ecology is that life on earth is possible only because of the recycling of matter. It is a rule, many ecologists believe, that we humans break at our peril and the peril of those yet to come.

2.1.3 Definition of Ecosystem

Ecologists invented the word "ecosystem", an abbreviated form of ecological system, to describe a network consisting of organisms, their environment, and all of the interactions that exist in a particular place. In short, an ***ecosystem*** is an interdependent and dynamic (ever-changing) biological, physical, and chemical system.

The biosphere is a global ecosystem. Because it is too complex to study, ecologists generally limit their view to smaller regions, setting up more manageable boundaries. For the sake of simplicity, an ecosystem might be a pond, a cornfield, a river, a field, a terrarium, or a small clearing in the forest. Accordingly, ecosystems vary considerably in complexity, too. Some may be quite simple – for example, a rock with lichens growing on it. Others, like the tropical rain forests, are quite complex. They contain an abundance of living organisms and a wide variety of species as well.

2.1.4 The Components of Ecosystem

All ecosystems are composed of two major parts: biotic and abiotic.

Biotic Factors The biotic components of an ecosystem are its living things which include producers, consumers, and decomposers.

Producers of an ecosystem mainly refer to green plants, and they also refer to photosynthetic bacteria and chemosynthetic bacteria. They are all called autotrophs. Green plants containing chlorophyll can convert carbon dioxide and hydrone to organic molecules through photosynthesis with the aid of sunlight. In this process, they release oxygen to the atmosphere. Photosynthesis can be written as follows:

$$6CO_2 + 6H_2O + E \longrightarrow C_6H_{12}O_6 + 6O_2$$

Photosynthetic bacteria and chemosynthetic bacteria can also bring inorganic molecules into organic molecules.

Consumers of ecosystem refer to those that directly or indirectly feed on green plants. They are called heterotrophs which include a variety of herbivores, carnivores, as well as some saprophytic or parasitic fungi.

Both plants and animals produce wastes and eventually die. This material forms a pool of dead organic matter known as *detritus*. In the ecosystem the organisms that use this detritus are known as *decomposers* which include some microscopic bacteria and fungi, such as flagellates, soil nematodes etc.

Abiotic Factors The abiotic, or non-living factors of an ecosystem are its physical and chemical components, for example, rainfall, temperature, sunlight, and nutrient supplies. Each of the earth's many organisms is finely tuned to its environment and operates within a range of chemical and physical conditions, namely, the range of tolerance. Although the range of tolerance is wide, most organisms thrive within a narrower range of conditions. It is called the optimum range. Outside this is the zone of physiological stress, where survival is possible, but difficult. Further outside the optimum range is the zone of intolerance, where an organism cannot survive. Fish, for example, generally tolerate a narrow range of water temperature. If the water cools below the lower limits of their range of tolerance, they will die or escape to warmer water. Water temperatures exceeding their upper limits of tolerance may also cause death or flight.

One of the problems with modern society is that it shifts environmental conditions, making regions hotter and drier. Such changes can make life more difficult, if not impossible, for other organisms. Changes in water temperature or the chemical composition of lakes, for instance, have dramatic impacts on fish and other aquatic organisms.

2.2 HOW DO ECOSYSTEMS WORK?

In this section we turn our attention to how ecosystems function. We will see how producers, consumers, and decomposers are related in an ecosystem and how energy and chemical nutrients flow through the biosphere.

2.2.1 Food Chains and Food Webs

In the biological world you are one of three things, a producer, a consumer or a decomposer. Producers are the organisms that support the entire living world through photosynthesis. Plants are the key producers of energy-rich organic materials. They are also called autotrophs, because they nourish themselves photosynthetically, that is, by using sunlight and atmospheric carbon dioxide to make the elements and materials they need to survive. Consumers feed on plants and other organisms and are called heterotrophs, because they are nourished by consuming other organisms.

Consumer organisms that feed exclusively on plants are called herbivores. Cattle, deer, and elk are examples. Those consumers that feed exclusively on other animals, such as the mountain lion, are *carnivores*. Those consumers that feed on both plants and animals, such as humans, bears, and raccoons, are omnivores.

Most of the decomposers are heterotrophs. They excrete the enzyme, a biological catalyst with protein characteristics, on the surface or inside the carcase and plant residues to decompose complex organic matter into simple inorganic compounds which can be re-absorbed by the producers.

The interconnections among producers, consumers, and decomposers are visible all around us. Mice living in and around our homes, for example, eat the seeds of domestic and wild plants and, in turn, are preyed on by cats and hawks.

The carcase and plant residues, called detritus, are the major food sources for the decomposers. The process of detritus decomposition is carried out by microscopic bacteria and fungi. Meanwhile, the decomposers derive energy and essential organic building blocks from detritus.

In the process they liberate carbon dioxide, water, and other nutrients needed by plants to make more plant material and maintain the perpetual cycle.

A series of organisms, each feeding on the preceding one, forms a food chain. Figure 2.2 illustrates familiar terrestrial and aquatic grazer food chains.

Figure 2.2 Examples of grazer food chains occurring on land and in water

(Source: Daniel D. Chiras, *Environmental Science: Action for a Sustainable Future*, The Benjamin/Cummings, 1991, P54)

Ecologists categorize consumers by their position in the food chain (Figure 2.3). For example, herbivores are called primary consumers, since they are the first organisms to consume the plants. Organisms that feed on primary consumers are secondary consumers, and so on.

The feeding level an organism occupies in a food chain is called the ***trophic level***. The first trophic level marks the beginning of the food chain and is made up of the producers or autotrophs. Primary consumers occupy the second trophic level, and secondary, tertiary, and quaternary consumers occupy the third, fourth, and fifth trophic levels, respectively. All consumers are heterotrophs. Figure 2.3 is an example of a food chain broken down into trophic levels.

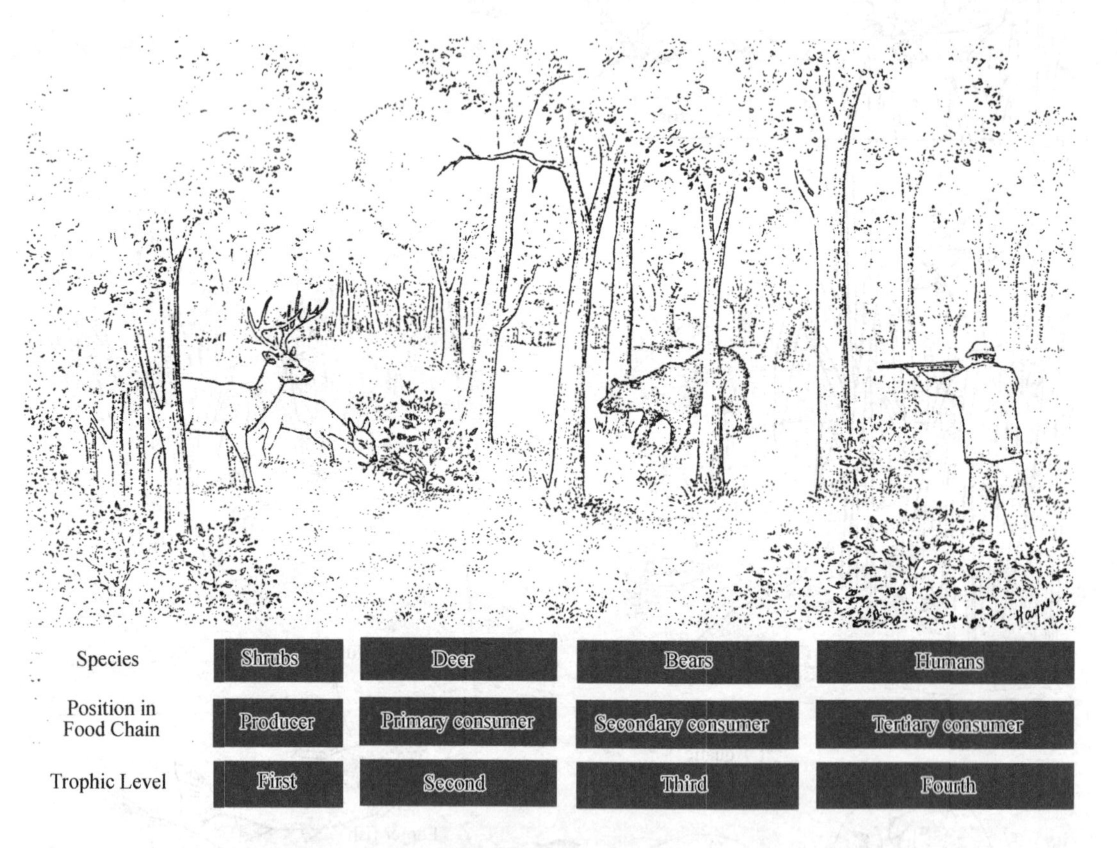

Figure 2.3 Food chain (showing trophic levels)

(Source: Daniel D. Chiras, *Environmental Science: Action for a Sustainable Future*, The Benjamin/Cummings, 1991, P57)

Food chains exist only on the pages of ecology texts; in reality, all food chains are woven into a much more complex network called a food web. The food web gives a complete picture of who eats whom (Figure 2.4).

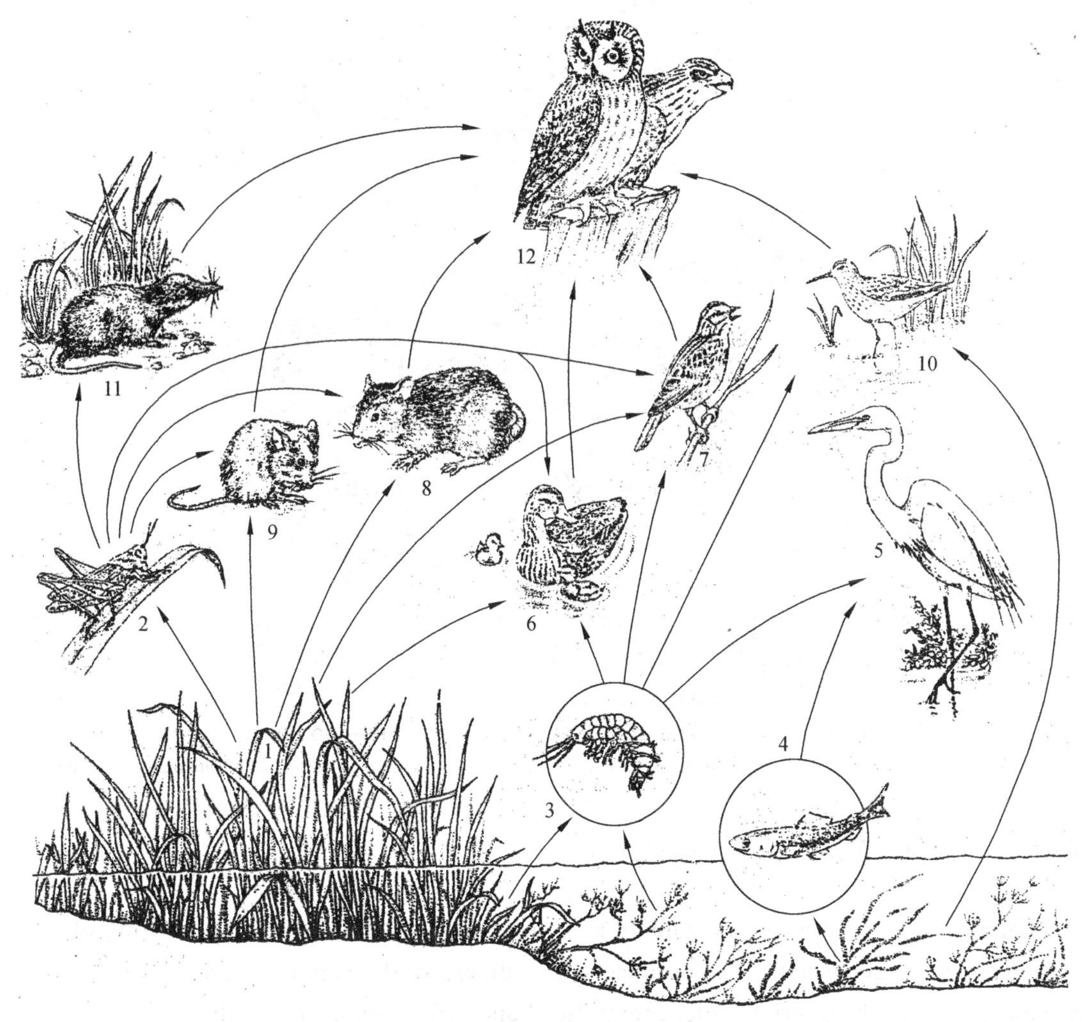

Figure 2.4 A simplified food web

1 – plants; 2 – grasshopper; 3 – invertebrates; 4 – fish; 5 – egret; 6 – mallard duck; 7 – sparrows; 8 – rat; 9 – harvest mouse; 10 – sandpipers; 11 – shrew; 12 – hawk and short-eared owl.

(Source: Daniel D. Chiras, *Environmental Science: Action for a Sustainable Future*, The Benjamin/Cummings, 1991, P58)

Trophic levels can be assigned in food webs just as in food chains; in a food web, however, many species occupy more than one trophic level. As illustrated in Figure 2.5, a grizzly bear feeding on berries and roots is acting as a primary consumer; it occupies the second trophic level. When feeding on marmots, an animal similar to the woodchuck, however, the grizzly is considered a secondary consumer and occupies the third trophic level. In other instances a grizzly may feed on insect-eating chipmunks. It thus occupies the fourth trophic level.

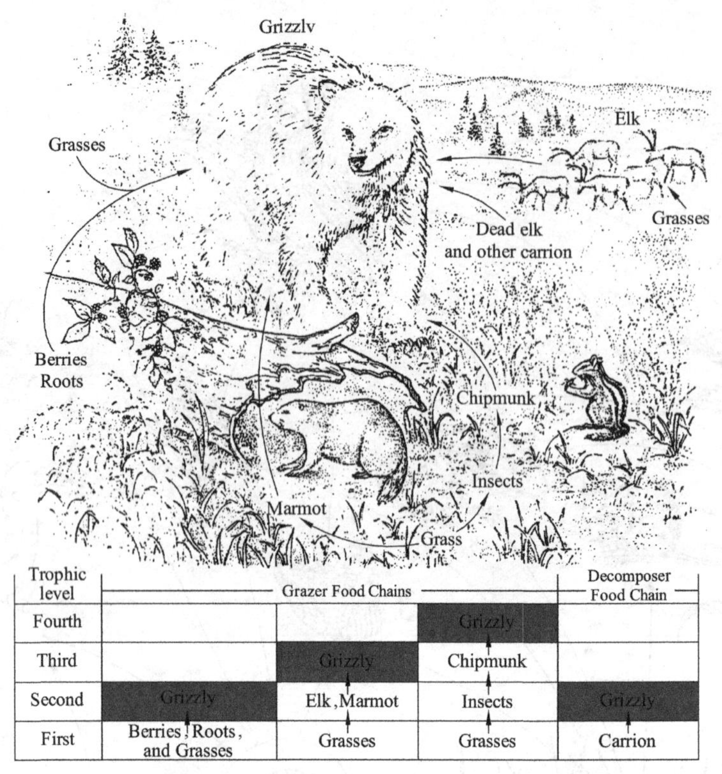

Trophic level	Grazer Food Chains			Decomposer Food Chain
Fourth			Grizzly	
Third		Grizzly	Chipmunk	
Second	Grizzly	Elk, Marmot	Insects	Grizzly
First	Berries, Roots, and Grasses	Grasses	Grasses	Carrion

Figure 2.5 The grizzly eats widely from the food web

(Source: Daniel D. Chiras, *Environmental Science: Action for a Sustainable Future*, The Benjamin/Cummings, 1991, P59)

2.2.2 Energy Flow in Ecosystems

The forementioned biotic components of an ecosystem are classified by the function of living organisms in the ecosystem. The producers, consumers, and decomposers are connected with one another through material cycle, energy flow, and information flow in the ecosystem.

Energy is not only the foundation of all life activities, but also the driving force of an ecosystem. Without energy transformation, ecosystem would not exist. In this section we will concentrate on the flow of energy through ecosystems.

All energy on earth originates from the sun as light energy. Plants are capable of tapping the sun, but they capture only 1% or 2% of the energy that the sun transmits to the earth (Figure 2.6). Still, on this small fraction of the sunlight is built the entire living world.

In ecosystems, all organisms are connected together through the food chain. The sun's energy is first captured by plants and stored in organic molecules which then pass through the food chains. In addition, plants incorporate a variety of inorganic materials such as nitrogen, phosphorus, and magnesium from the soil. These chemical nutrients become part of the plant's living matter.

Energy flow is only in one direction. When the green plant is consumed by animals, the energy flows from plants to herbivores (primary consumers). If the herbivores are eaten by carnivores (second consumers or third consumers), the energy flows to the carnivores. At last, the detritus are decomposed by various microorganisms and the complex compound organic matters are decompounded to simple inorganic matters, such as carbon dioxide, water, and sulfur dioxide. The energy is eventually returned to the environment. So, the energy flows from the sun to the producers, then to the decomposers. Meanwhile, the energy flows from the first trophic level to the second, third, and fourth, respectively. Figure 2.7 shows the energy flow in ecosystems.

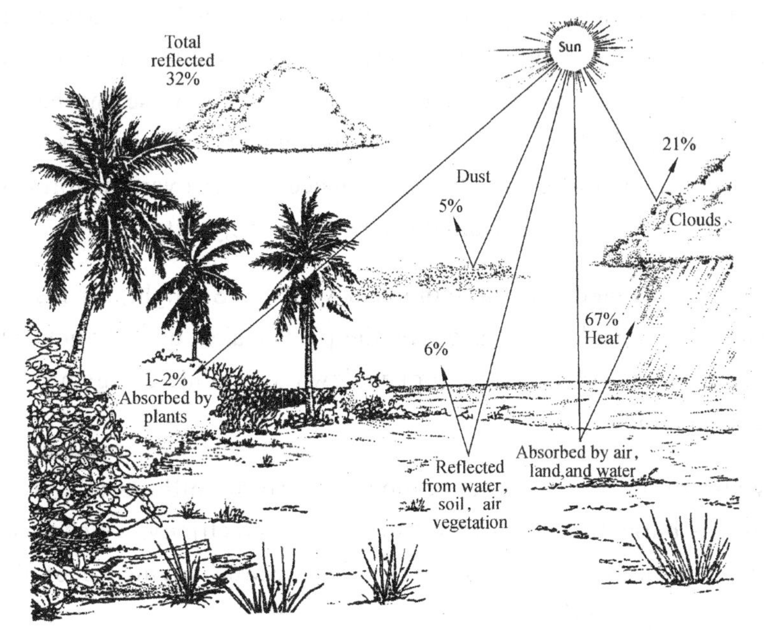

Figure 2.6 Distribution of incoming solar radiation

(Source: Daniel D. Chiras, *Environmental Science: Action for a Sustainable Future*, The Benjamin/Cummings, 1991, P61)

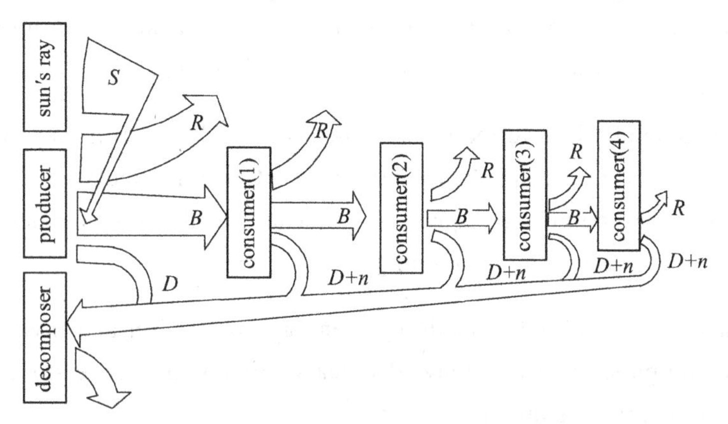

Figure 2.7 The way of energy flow in ecosystem

S-sun's ray; R-breathing dissipative energy; B-existing biomass; D-detritus; D+n-detritus and dejecta

It is necessary to note that the flow and transformation of energy must obey the first and the second law of thermodynamics.

The First Law The first law of thermodynamics is often called the law of conservation of energy. It states that energy can be neither created nor destroyed but can only be transformed from one form to another.

Let's look at a familiar example. The gasoline in your car contains an enormous amount of potential energy that, when released in combustion, speeds your car down the highway. As you drive along, the gas gauge shows how much gasoline has been used. Contrary to what you may think, you have not destroyed the energy. Instead, you have converted it into other forms – electricity to run your radio and windshield wipers, heat to keep you warm and defrost your windows, light to show you where you are going, and, of course, mechanical movements that propel your car along the highway. Careful measurement of the amount of energy your car is consuming and the amount of energy being produced in these various forms shows that the two are equal. In simple terms, energy input is equal to energy output.

Many energy conversions take place in biological systems. Sunlight, for instance, is trapped by plants and stored in organic food materials the plant makes during photosynthesis. These molecules are ingested by herbivores and broken down to provide energy needed to perform a variety of cellular functions.

Modern society also relies on energy conversions to perform millions of activities every day. Coal, for example, is burned in power plants to generate electricity, which is used, in turn, to power light bulbs, neon signs, and electric motors. Surely, evidence of the first law is everywhere, but how important is it?

The first law helps us understand that energy exists in many forms. Coal, for example, is a type of potential energy. It can be converted into heat, light, and electricity – all forms of kinetic energy. Of these forms, electricity is the most widely used one.

The first law is also popularly referred to as the no-free-lunch law. It means that in all systems, energy output can never be greater than energy input. Careful calculations always show that energy input and energy output are equal. The first law of thermodynamics is often said to mean that you can't get something for nothing.

The Second Law The first law deals with energy conversion and involves quantities – energy inputs and outputs. The second law also deals with energy conversion, but it involves a different aspect of energy – its quality.

More specifically, the second law of thermodynamics explains what happens to energy quality when energy changes from one form to another. The law simply states that energy is "degraded" during such conversions.

Another way of saying it is that energy goes from a concentrated to a less concentrated form during a transformation. For example, when gasoline is burned in an automobile, it is converted from a very concentrated form to much less concentrated heat, which is no longer available for useful work. Concentrated energy forms are said to have a great deal of available work. The less concentrated forms have lower capacity and are said to have less available work. Consider some examples. Oil and natural gas are high-quality energy sources. They are concentrated fuels that, when burned, can be put to good use. Most heat, however, is a low-quality energy source. It is quickly diluted or dispersed and is a less useful form of energy to power machinery. All heat produced on earth eventually dissipates into space and is lost to us.

The second law of thermodynamics has many implications for our lives. It tells us that when we burn fossil fuels, our supply of highly concentrated energy shrinks. It tells us also that we cannot recycle high-quality energy, because when it is burned it is dissipated into heat and lost into space. And it warns us not to waste this precious resource; the source we now tap is all that we have.

Biomass and Ecological Pyramids The laws of thermodynamics go much further. They rule the living world, from the tiniest bacterium to the largest whale. They limit ecosystems, as you will soon see, and can be used to guide us along a sustainable pathway. Let us look at the implications of these laws from an ecological perspective. We will begin with the concept of biomass.

Biomass is organic matter created by plants and other photosynthetic organisms and passed up the food chain. Because organisms vary considerably in their water content, water must be excluded when determining biomass. *Biomass* is, therefore, the dry weight of living things and can be measured for each trophic level. To sample the biomass of plants, for example, a single square meter of vegetation, roots and all, would be removed dried, and weighed. The result is the plant's biomass. The biomass at the first trophic level in almost all ecosystems represents a large amount of potential (chemical) energy and tissue-building materials for the second trophic level. Studies of ecosystems show, however, that not all of the biomass of the first trophic level is converted into biomass in the second trophic level. In other words, not all plant matter becomes animal matter. Several reasons can be given for this fact. The first is that only a small part of the plant matter in any given ecosystem is eaten by organisms of the next higher level. Second, not all of the biomass eaten by the herbivores is digested; some passes through the gastrointestinal tract unchanged and is excreted. Third, most of what is digested is broken down into carbon dioxide, water, and energy used to move about, to breathe, and to maintain body temperature. The second law of thermodynamics tells us that this energy is converted into heat, which is dissipated into the environment and is eventually lost into space.

Because of these factors the biomass of the second trophic level is greatly reduced. Ecologists once thought that 10% of the biomass at one trophic level was transferred to the next. They called this *the ten percent rule*. Further research showed, however, that the amount of biomass transferred from one level to the next varies from 5% to 20% in different food chains. Graphically represented, biomass at different trophic levels forms the pyramid of biomass (Figure 2.8).

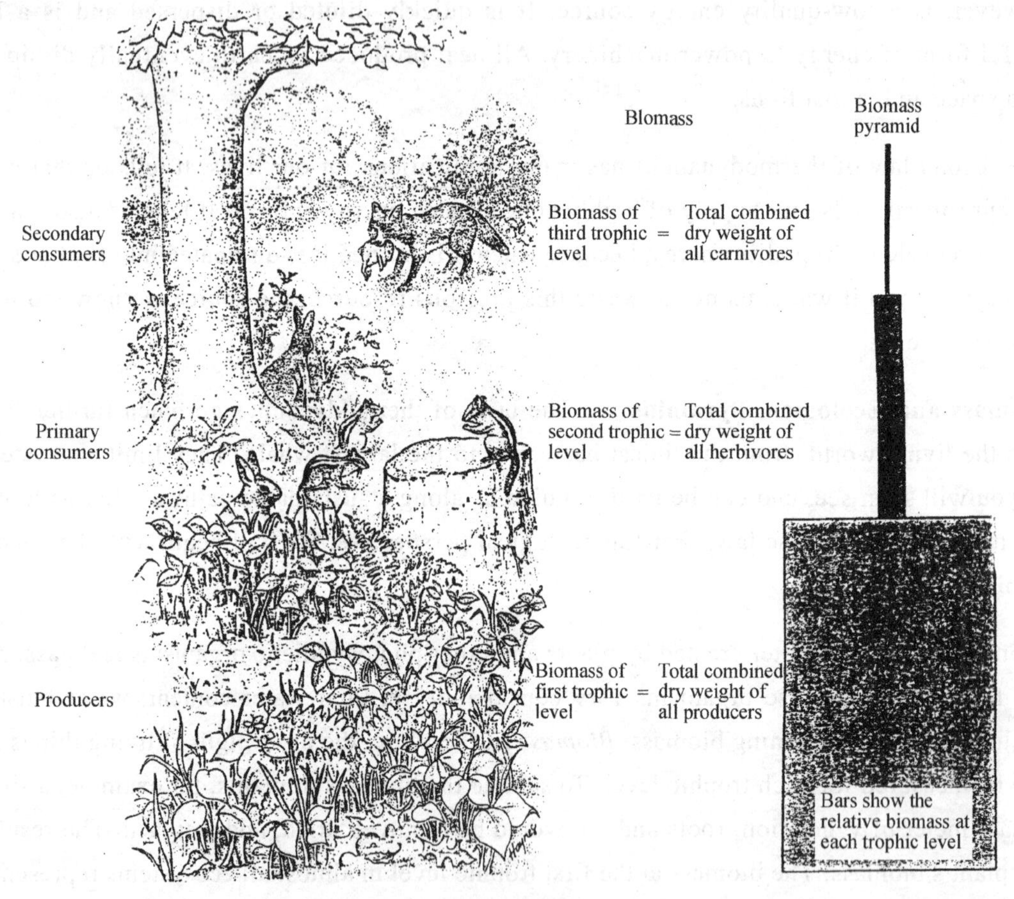

Figure 2.8 Biomass pyramid

(Source: Daniel D. Chiras, *Environmental Science: Action for a Sustainable Future*, The Benjamin/Cummings, 1991, P64)

Biomass is the substance that makes up living things. The chemical bonds that hold the organic compounds of biomass together contain enormous amounts of stored, or potential, energy. This energy can be released when organic matter burns or when it is broken down by cells in plants, animals, and microorganisms. If the energy content of biomass is graphed as the biomass was, it, too, forms a pyramid, called *the pyramid of energy* (Figure 2.9).

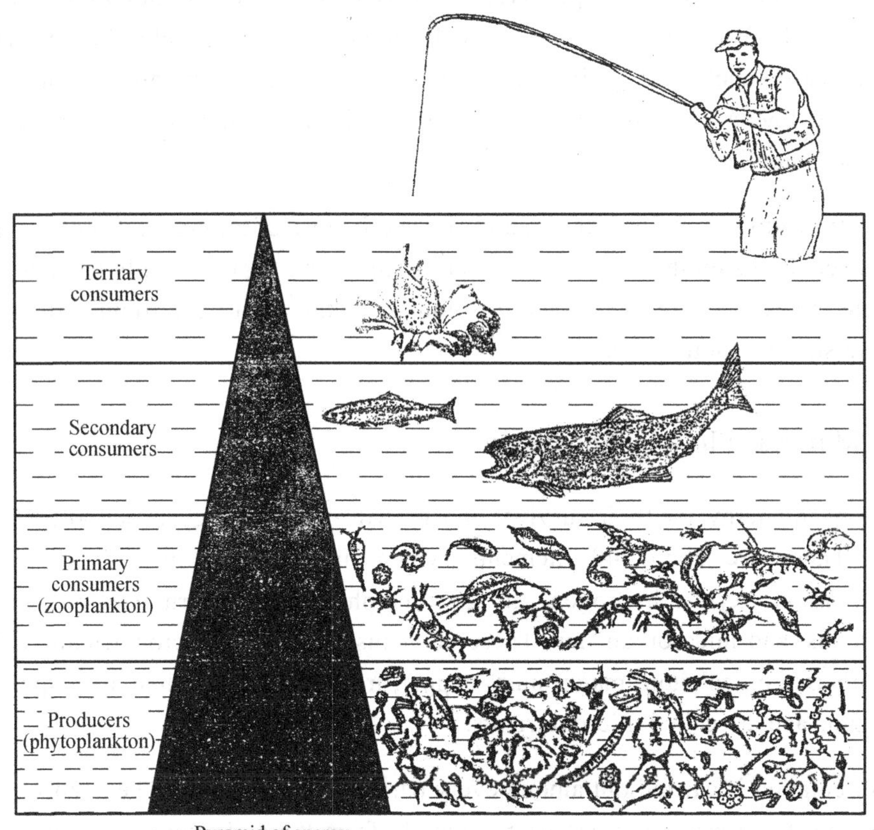

Figure 2.9 Pyramid of energy

(Source: Daniel D. Chiras, *Environmental Science: Action for a Sustainable Future*, The Benjamin/Cummings, 1991, P65)

What are the implications of the ecological pyramids? As you have learned earlier, biomass decreases in the upper trophic levels of a food chain. Thus, fewer organisms can be supported at these higher levels. Graphically represented this forms a pyramid of numbers.

Over twice as many herbivores can be supported in a grassland biome as carnivores. Carnivores, in fact, always have the smallest populations in ecosystems. Thus, they are often the first endangered by human activities that disrupt the food chain. A reduction in biomass produced at the first trophic level, for instance, can have profound impacts on carnivores because it decreases prey populations. Take away its prey and the carnivore is left to starve or enter into fierce competition with its own kind for food.

Another implication of ecological pyramids is that more organisms can be supported in an ecosystem if they can feed at lower trophic levels. In human terms, then, consuming meat is much more wasteful of solar energy than eating plants, calorie for calorie. If 20,000 kilocalories worth of corn were fed to a steer, for example, 2000 kilocalories of beef would be produced (using a 10% conversion). This would feed only one person (assuming that a person can survive on 2,000 kilocalories per day). If the 20,000 calories of corn were eaten directly, however, it would feed ten people for a day. In improving food supply, then, it makes more sense to increase supplies of grain rather than meat, which is generally the approach most countries take. Encouraging people to eat lower on the food chain could also help increase available food supplies.

The third implication of ecological pyramids is that the loss of biomass from one trophic level to the next sets limits on the length of the food chain. Food chains usually have no more than four trophic levels, because the amount of biomass at the top of the trophic structure is not sufficient to support another level.

2.2.3 Material Flow in Ecosystems

Organisms need energy for their survival and growth, while energy is supported by various materials. There are about 40 elements which are essential to life. Five of these elements – carbon, oxygen, hydrogen, nitrogen, and phosphorus – form 97% of the mass of all plants, animals, and microorganisms. These elements are called macronutrients. Others such as sulfur, potassium, calcium, magnesium, copper, iron, zinc, and iodine are required in only small amounts and are called micronutrients.

Material circulation is carried out among every part of the ecosystem. The supply of nutrients other than CO_2 to an ecosystem comes principally from the soil, and also from the air, rain, snow and dust,to a smaller extent. Nutrients are cycled in such a way that they are either incorporated into plants and animals, or made available for plant uptake by the decomposition of dead plant and animal remains. The pathway from sources to sinks and back to sources are termed material cycles, or material flow.

Material cycle is accompanied by energy flow inside ecosystems, but there is a fundamental difference. Energy flow is in only one direction while material flow is cyclical. All of the energy that enters a food chain is eventually lost as heat. The low-quality energy is dissipated into space and cannot be reused. But nutrients cannot disappear. They can be reused by plants.

Material cycles usually include biological, geological, and chemical cycles, so material cycles that move nutrients through the biosphere are also called biogeochemical cycles, or nutrient cycles.

This section examines four nutrient cycles: water, carbon, nitrogen, and phosphorus.

The Hydrological Cycle The global recycle of water is the hydrological cycle, or water cycle. It runs day and night, free of charge, busily collecting, purifying, and distributing water that serves a multitude of purposes along its path. This cycle includes the precipitation of water from clouds, infiltration into the ground or runoff into surface watercourses, followed by evaporation and transpiration of the water back into the atmosphere (See Figure 2.10).

Evaperation and transpiration are the two ways water reenters the atmosphere, from free water surfaces or by plants. The same meteorological factors that influence evaporation are at work in the transpiration process: solar radiation, ambient air temperature, humidity, and wind speed, as well as the amount of soil moisture available to the plants – these all impact the rate of transpiration. Because evaporation and transpiration are so difficult to measure separately, they are often combined into a single term, evapotranspiration.

Precipitation is the term applied to all forms of moisture originating in the atmosphere and falling to the ground (e.g., rain, sleet, and snow). Precipitation is measured with gauges that record in centimetres of water. The depth of precipitation over a given region is often useful in estimating the availability of water.

Water suspended in the clouds, except in polluted regions, is nearly pure. The reason for the purity of atmospheric moisture is that when water molecules evaporate, they leave behind dissolved impurities.

In the atmosphere, water is suspended as fine droplets. The amount of moisture the air can hold depends on air temperature. The warmer the air is, the more moisture it can contain. Atmospheric moisture content can be expressed as absolute humidity – the number of grams of water in a kilogram of dry air – or as relative humidity, the more common measurement. Relative humidity measures how much moisture is present in the air compared with how much it could hold if fully saturated at a particular air temperature. At a relative humidity of 50%, for example, air has 50% of the water vapor it can hold at that temperature. If the relative humidity is 100%, the air is saturated.

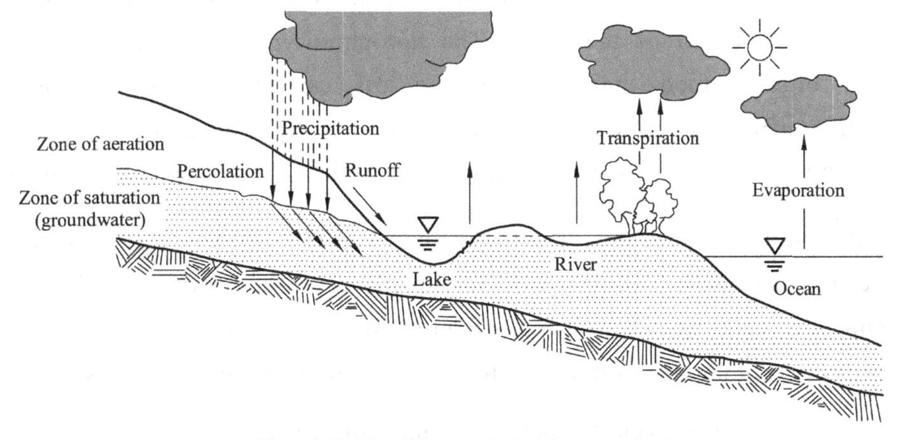

Figure 2.10 The hydrological cycle

(Source: P. Aarne Vesilind and Susan M. Morgan, *Introduction to Environmental Engineering*, Thomson Brooks Cole, 2004, P197)

When moisture exceeds saturation, clouds, mist, and fog form. Clouds, for example, form when moist air is raised by mountain ranges, when cold-air masses come in contact with moisture-laden air, or as warm air rises to cooler levels. For rain to form, air must contain small particles, known as condensation nuclei, on which the water vapor collects. Condensation nuclei may be salts from the sea, dusts, or particulates from factories, power plants, and vehicles. Over a million fine water droplets must come together to make a single drop of rain. If the air temperature is below freezing, the water droplets may form small ice crystals that coalesce into snowflakes.

Clouds move about on the winds, generated by solar energy, and deposit their moisture throughout the globe as rain, drizzle, snow, hail, or sleet. Precipitation returns water to lakes, rivers, oceans, and land from which it comes, thus completing the hydrological cycle. Water that falls on the land may evaporate again or may flow into lakes, rivers, streams, or groundwater, eventually returning to the ocean.

The Carbon Cycle Carbon is the essential component of all organic matter. The ultimate source of carbon for organic matter is carbon dioxide, converted to organic matter in photosynthesis. In nature, carbon dioxide consists in the atmosphere or dissolves in the water. Carbon cycle is one of the most important nutrient cycles. It is responsible for the recycling of carbon dioxide given off by all living things.

The carbon cycle consists of two halves. The first half is photosynthesis. This is the phase during which carbon dioxide is taken up by plants and algae and converted to food molecules with the aid of sunlight. The chief products of photosynthesis are oxygen and organic molecules which are distributed through the food chain and make up the tissues of living matter. Half the oxygen in the atmosphere is replenished each year by plants and algae. Both oxygen and organic molecules are essential nutrients for non-photosynthetic organisms, which constitute the second half of the cycle. The non-photosynthetic organisms consume the oxygen and organic food molecules, giving off carbon dioxide to complete this ever-continuing cycle of mutual dependence. Thus, virtually all animals exists on the earth because of plants. Plants, in turn, survive thanks to carbon dioxide released by animals.

A simplified version of the carbon cycle is shown in Figure 2.11. Atmospheric carbon dioxide enters terrestrial and aquatic ecosystems. Within plants, carbon dioxide molecules react to form organic molecules such as glucose. Energy for these reactions comes from sunlight through the process of photosynthesis. Photosynthesis can be written as follows:

carbon dioxide + water + sunlight→glucose+oxygen

$$6CO_2 + 6H_2O + E \rightarrow C_6H_{12}O_6 + 6O_2$$

As illustrated in Figure 2.11, the organic molecules produced during photosynthesis are passed through the food web. Thus, carbon flows from the atmosphere into and through the organismic phase of the cycle, travelling through food chains.

Carbon returns to the atmosphere in several ways. In plants and animals, for instance, some of the organic molecules are broken down to generate cellular energy. This process, called cellular respiration, results in the production of usable energy, heat, carbon dioxide, and water. It can be written as follows:

glucose +oxygen→carbon dioxide+water+energy

$$C_6H_{12}O_6 + 6O_2 \rightarrow 6CO_2 + 6H_2O + E$$

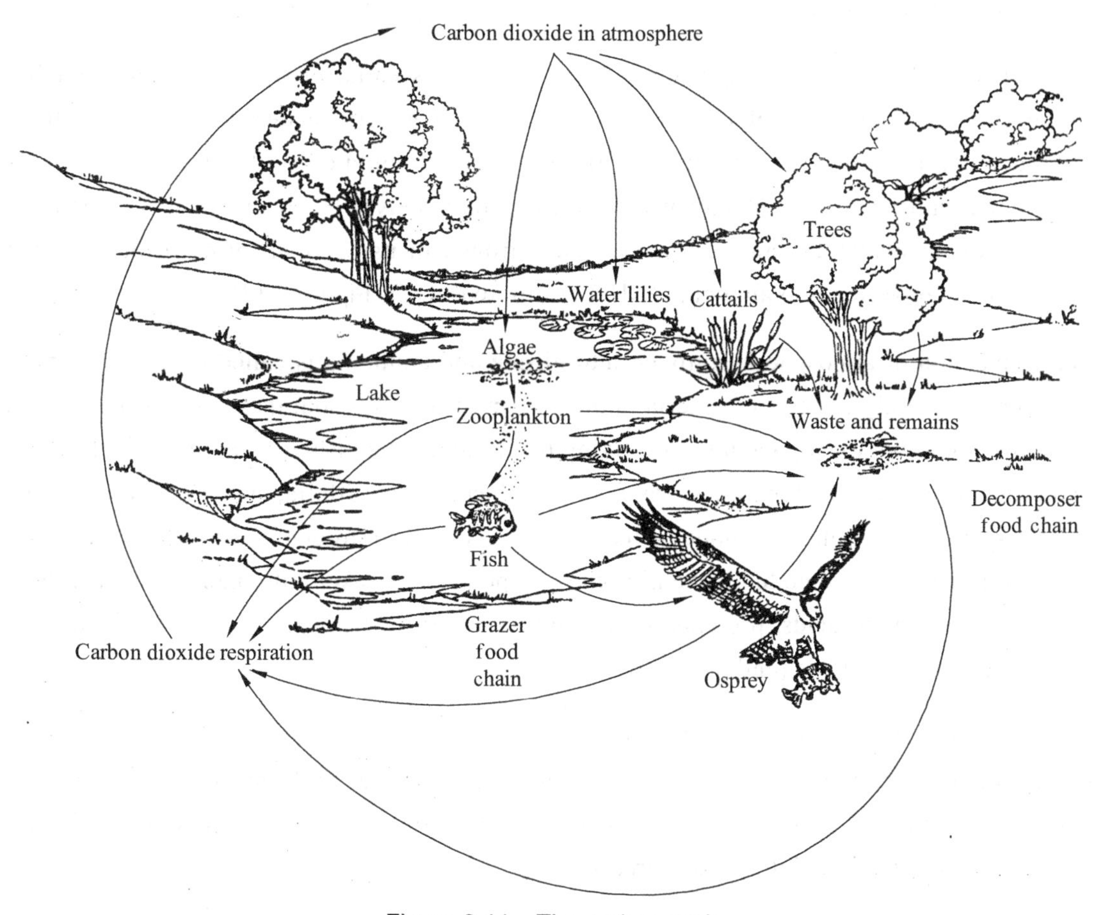

Figure 2.11 The carbon cycle

(Source: Daniel D. Chiras, *Environmental Science: Action for a Sustainable Future*, The Benjamin/Cummings, 1991, P68)

This reaction is the reverse of photosynthesis. The carbon dioxide gas released during cellular respiration reenters the environmental phase of the cycle for reuse.

Carbon can also return to the atmosphere through the activities of groups of bacteria and fungi, which use dead organic matter as their food source. They thus oxidize the dead material, either directly or in a number of stages, with CO_2 and H_2O as end products, and complete the cycle. Carbon also returns when plant materials are burned by natural causes such as lightning and forest fires or as a result of human activities such as combustion of wood and coal or deliberately set fires.

Humans intervene in the global carbon cycle in two major ways: 1) by removing forests and vegetation that use atmospheric carbon dioxide to make organic molecules, and 2) by liberating carbon dioxide during the combustion of coal, oil, and natural gas – carbon sources that were once isolated deep beneath the earth's surface. Combined, such activities have increased the global carbon dioxide concentrations. Further increases could have a devastating effect on global climate and life.

The Nitrogen Cycle Nitrogen forms part of many essential organic molecules, notably amino acids (the building blocks of proteins) and the genetic materials RNA and DNA. Without a continuous supply of nitrogen, life on the earth would cease. Fortunately for plants and animals, nitrogen is an abundant element; approximately 79% of the air is nitrogen gas (N_2). However, plants and animals cannot use nitrogen in its original form. To be usable it must first be converted into ammonia (NH_3) or nitrate (NO_3^-) (Figure 2.12).

The conversion of atmospheric nitrogen into nitrate and ammonia is called *nitrogen fixation*, and it occurs mainly in certain bacteria in the soil and water. Without these organisms, life, as we know, could not exist. One nitrogen-fixing bacterium, called Rhizobium, invades the roots of a group of plants called legumes, which include beans, peas, alfalfa, clover, and others. The roots respond by forming tiny nodules that serve as sites for nitrogen fixation. Formed either in the soil or in the root nodules, the nitrogen compounds are taken into plants and used there to synthesized amino acids, proteins, DNA, and RNA. Animals, in turn, receive the nitrogen they need by eating plants (and other animals).

Nitrogen is returned to the soil by the decay of detritus (Figure 2.12). Within the soil certain species of bacteria and fungi decompose the nitrogen-rich wastes from plants and animals. Nitrogen is released in the form of ammonia and ammonium salts. Ammonia is further converted into nitrites and then into nitrates.

The ammonium salts, nitrates, and nitrites may all be incorporated by the roots of plants and reused. Some nitrite is converted into a gas, nitrous oxide (N_2O), and released into the atmosphere.

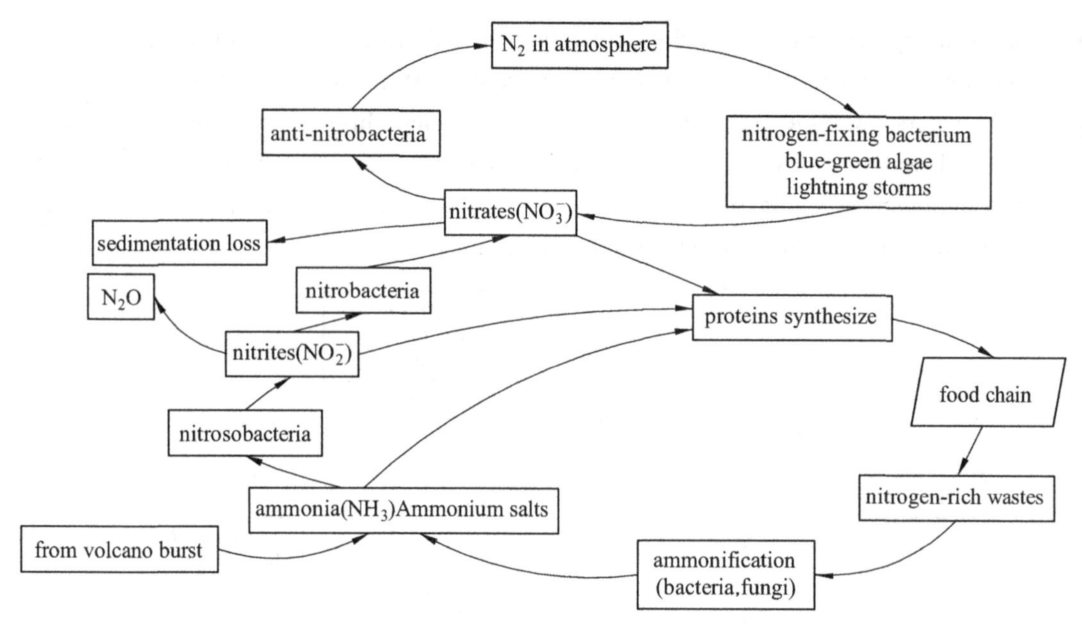

Figure 2.12 The nitrogen cycle

In various forms nitrogen travels from air to soil to plant to animal and then back to soil and atmosphere in a never-ending cycle.

Human farming practices greatly affect the nitrogen cycle. For example, some crops such as corn absorb large amounts of nitrogen from the soil; if nitrogen is not replaced in alternate years, the soil may become unproductive. Farmers may replace soil nitrogen by applying artificial fertilizers that contain nitrates. Frequently used in excess, these nitrates may wash into streams, causing excessive nuisance growth of algae and macrophytic plants.

The phosphorus Cycle Phosphorus, found in living organisms as phosphate (PO_4^{3-}), is an important part of RNA and DNA. Phosphorus is also found in fats (phospholipids) in cell membranes.

Phosphorus mainly comes from phosphate ores, bird droppings, and natural animal fossils. Phosphate is slowly dissolved from rocks by rain and melting snow and is carried to waterways. Dissolved phosphates are incorporated by plants and then passed to animals in the food web. Phosphorus reenters the environment in at least two ways: 1) some is excreted directly by animals, and 2) some is returned when detritus decays (Figure 2.13).

Each year large quantities of phosphate are washed into the oceans, where much of it settles to the bottom and is incorporated into marine sediments. Sediments may release some of the phosphate needed by aquatic organisms. The rest may be buried and taken out of circulation.

The input of phosphorus from human activity can be far greater than from natural sources. Domestic sewage contains phosphorus in feces and from commercial detergents, in which phosphates are used (as wetting agents), although the latter's contribution has been greatly reduced in many countries, following legislation. Runoff from agricultural areas which have received fertilizers can be another important source of phosphorus. Therefore, in many polluted waters, soluble phosphorus can reach much higher concentrations than in non-polluted waters. This readily available phosphorus can often lead to the growth of nuisance organisms such as filamentous algae, which can cause taste and odor problems in water supplies and clog filters in water treatment plants.

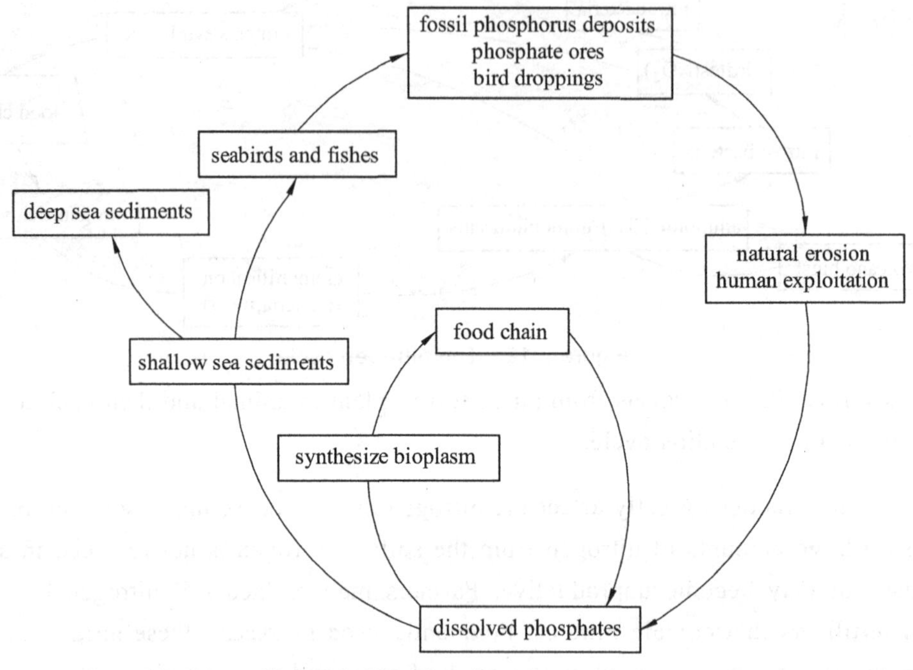

Figure 2.13 The phosphorus cycle

2.2.4 Infomation Flow in Ecosystems

The function of an ecosystem not only incarnates energy flow and substance repetition, but represents that there is an information connection (or information transfer) among biologic components in the system, which is termed "Information Flow" customarily. Various kinds of information existing in an ecosystem combine biologic components into a whole ecosystem. The information in an ecosystem can be principally divided into physical information and chemical information.

Physical information The physical information in an ecosystem consists of light, sound (such as twitter, animal roar), and colour. The behavior of creatural courtship and threat as well as alarm relates to physical information. The physical information includes biologic behavior information. Birds in the season of reproduction, for instance, usually company with bright-coloured feather or other oddity decoration or sweetness twitter. Their strong suit completely unveils in the process of courtship. Male peafowl opens its beautiful feather, which can impels female peafowl to approach him, and female peafowl puts a pose for accepting caress, and the performing action of male peafowl can compels them to couple rapidly. Insects may judge the food (nectar) according to flower colour, for another instance. Fishes are used to taking light as their food information.

Chemical information Some substances in biological metabolism excretion, such as vitamin, growth hormone, bacteriophage and sex hormone, belong to chemical matter of transferring information. The amount of these substances is very small, but they may realize information transfer in ecosystems and contact organisms among the population or inside it. Some organisms may be restrained and some organisms attract one another by the chemical matter. For example, many female insects release sex hormone as sexual temptation to male insects. Weasels have a scent gland releasing unpleasant smell which can prevent enemy pursuit, and also be conducive to capturing food. Ants leave a chemical trial through their secretions for the latter to follow. Nicotine and other alkaloids in tobacco can make aphides numbness. The leaf surface of a walnut tree is able to produce a matter of growth hormone which can inhibit the growth of other shrubs and herbaceous plants but avail the growth of the walnut tree itself, when it is washed into the soil. These are chemical information for their self-protection.

Although there are many difficulties in the studies on information flow in ecosystems, due to the limitation of science technology level, the information relations among organisms can adjust and affect ecosystems quite obviously, especially the chemical information. For instance, wolves use their stale to mark the moving path, and they usually use stub and tree as "smell station". Sometimes, a group of wolves micturate in turn at the same marker. This smell station forms good-sized ice lump in winter, through which people can obtain the information of the amount of the wolf troop by analyzing ice lump.

2.3 ECOSYSTEM BALANCE AND IMBALANCE

In 1884, the water hyacinth was introduced into Florida from South America. The hyacinth remained, for a while, simply a beautiful ornamental flowering plant in a private pond. Unfortunately, the plant accidentally entered the waterways of Florida. In these nutrient-rich waters it spread like cancer throughout the canals and rivers that lace the state. Aided by a remarkable ability to reproduce – 10 plants can multiply to 600,000 in just eight months – the hyacinth now clogs waterways throughout Florida, choking out native species and making navigation impossible in some areas.

From Florida the plant has spread throughout most of the southern United States. Today nearly 800,000 hectares of rivers and lakes from Florida to California are chortled with choked subface mats of hyacinths. Florida, Louisiana, and Texas, where the problem is most severe, spend nearly $11 million a year to relieve the stranglehold these plants have on their waterways.

For most of its existence humankind has worried about the ways in which nature affects people's lives. As this example points out, however, the danger may not be what the nature will do to us, but what we will do to the nature (and ourselves) by unleashing some of its forces.

The story of the water hyacinth is a lesson in ecological balance, or more correctly, imbalance, a lesson in how an ecosystem can be thrown out of kilter by humankind's zeal to fashion the environment to its liking.

2.3.1 Ecosystem Stability Defined

In a certain period of time, the producers, consumers, and decomposers of an ecosystem constantly carry out material cycle, energy flow, and information flow and keep a kind of dynamic balance, which is called the dynamic equilibrium, or the ecological balance,or a steady state. In other words, it is a state of stability. To understand this term, suppose that you have studied a mature forest ecosystem near your home each spring for an extended period, say, 20 years. If the system were stable, you would find that 1) the total number of species was fairly constant from year to year, 2) the same species were present each year, and 3) the population size of each species was approximately the same from year to year. This system is stable.

Stability doesn't mean that all of the parts of an ecosystem operate in perfect harmony. Ecosystem stability or balance, in fact, is often achieved through competition and apparent conflict: animals competing for a limited food supply, disease organisms killing off the weak, and predators feeding on prey. The net result, or the more or less constant condition, is what ecologists refer to as stability.

Ecosystems, then, are complex self-regulating systems. As in all such systems, a change in one variable results in a corresponding change in another, bringing the system back into alignment. The chief form of regulation is negative feedback. The ability to bounce back from small changes is sometimes called resilience.

Perhaps the key to understanding stability is to realize that it does not mean that conditions are absolutely constant. Just as your blood sugar levels vary slightly from hour to hour, so do conditions within an ecosystem. A stable ecosystem, however, can weather such variation. That's because species operate within a range of conditions – the range of tolerance.

When conditions are shifted out of the range of tolerance, an ecosystem may begin to deteriorate, losing individual species that are vital links in food webs.

2.3.2 What Keeps Ecosystems Stable?

Nature is a balancing act of growth and decline, predator and prey, sickness and health; the secret of living systems is achieving a balance and maintaining it. Ecosystem health, like your own health, is dependent on this precarious balance.

Population Growth and Environmental Resistance Balance, or stability, in an ecosystem is the result of opposing forces that constantly work to regulate the size of populations. These forces can be broken down into two groups: factors that tend to increase population size, or growth factors, and those that tend to decrease it, called reduction factors, terms coined by ecologists to describe positive and negative feedback mechanisms that help regulate populations. Growth and reduction factors can be biotic or abiotic.

At any given moment population size is determined by the interplay of these factors. Since ecosystems contain many species, the entire ecosystem balance can be crudely related to the sum of the individual population balances.

The biotic factors that stimulate population growth include the ability to produce many offspring, to adapt to new or changing environments, to migrate into new territories, to compete with other species, to blend into the environment, to defend against enemies, and to find food. Certain favorable abiotic conditions also tend to increase population size. Favorable light, temperature, and rainfall, for example, all promote maximum plant growth and, because animals depend on plants, often promote increases in animal populations.

Opposing the positive influence on growth are a host of abiotic and biotic factors. Ecologists describe these factors collectively as environmental resistance. Predators, disease, parasites, and competition by other species all effectively reduce population size, as do unfavorable weather and lack of food and water. In the case mentioned above, the water hyacinth faces little environmental resistance. Even plant-eating native fish are no match for its reproductive success.

Before the invention of the plow, unbalanced ecosystems were a rare thing. Expansion of agriculture, urban growth, and industrial development changed the picture entirely, making balanced ecosystems increasingly difficult to find in many parts of the world. We will focus on a tiny portion of that ecosystem to illustrate how nature ensures its harmony.

Living in the grass-covered Flint Hills is a mouselike rodent called the prairie vole. Its population size depends on many factors such as light, rainfall, available food supply, predation by coyotes and hawks, temperature, and disease.

The prairie vole has a high reproductive capacity. When raised in the laboratory, under optimum conditions, a pair of voles become remarkably prolific, producing a litter of seven pups every three weeks, month after month, for several years. The optimum conditions in a laboratory are low temperature (slightly above freezing), long days (14 bouts of light a day), and plenty of water and food. Under these conditions a female will give birth to a litter and mate a few hours later. Her second litter is born about the time she weans her first. This can go on, in the laboratory, for several years. For the captive-raised vole, motherhood is no picnic.

Fortunately, in the wild, optimum conditions never completely coincide. In the summer, for instance, when food, water, and day length are optimal, the temperature is too warm. Reproduction occurs at a much more reasonable pace. In the winter, when the temperature is just right for breeding, the days are short, and the food supply is reduced. As a result, the population balance is maintained.

Resisting Change The key word in ecosystems is stability or balance. As all of us can attest, to preserve balance, it is easiest to resist change in the first place. In an ecosystem this is called inertia. Small changes in water chemistry, for example, may have no effect on aquatic organisms. As a result, the aquatic system resists change. If change does occur, ecosystems may "bounce back" or recover rapidly; this is called resilience. This section discusses the phenomenon of resilience.

In the living world, change comes from shifts in growth and reduction factors. The introduction of new predators, a shortage of food, low rainfall, or unfavorable temperature, for example, all tend to decrease population size. Other factors, such as an abundance of food, may cause explosive growth in populations.

Changes in abiotic and biotic conditions occur with great regularity in ecosystems. Minor fluctuations are of little consequence, because species have evolved numerous mechanisms and can either resist change or recover quickly. Prairie voles may increase their reproductive rate when a cold winter kills off a larger than normal number of their population; tropical monkeys may exploit new food sources when a rainy season fails to materialize and traditional food supplies are inadequate; wolves may migrate into new territory if crowding occurs.

In many ways nature is a series of checks and balances that preserve the integrity of the whole. These checks and balances help minimize human impact, too. Sewage dumped into a stream, for example, adds organic and inorganic chemicals to the water (Figure 2.14). The organic molecules are consumed by naturally occurring bacteria whose population is normally low. The number of bacteria increases. Since bacteria use up oxygen when they consume organic materials, the level of dissolved oxygen in the stream usually drops. This decline kills fish and other organisms or forces them to migrate to new areas. In time, though, the stream will return to normal if further spills are prevented. The bacteria that proliferated after the spill will perish as the level of organic pollutants falls off. The levels of dissolved oxygen will also return to normal. Fish will return. This is an example of resilience.

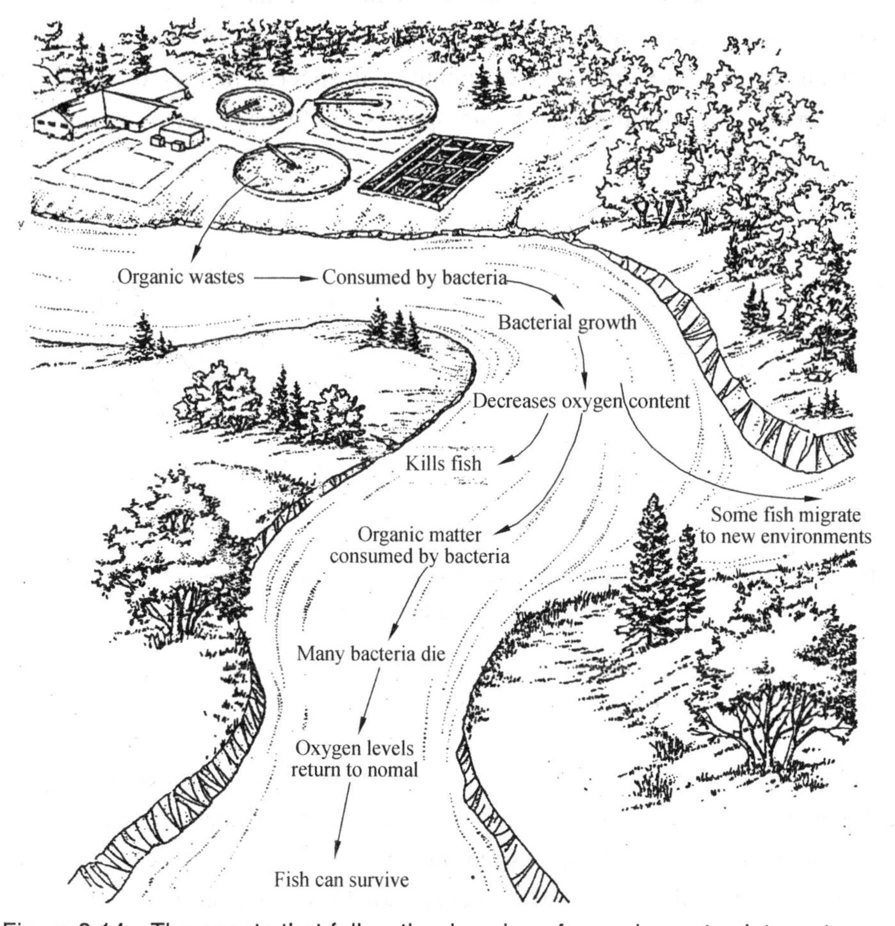

Figure 2.14 The events that follow the dumping of organic wastes into a stream

(Source: Daniel D. Chiras, *Environmental Science: Action for a Sustainable Future*, The Benjamin/Cummings, 1991, P79)

Species Diversity and Stability Ecosystem stability is significantly affected by species diversity, which, roughly speaking, is a measure of the number of species in a community. The relationship between species diversity and latitude is found in virtually all groups. Latitude, therefore, is an important factor affecting species diversity, but it is not the only one.

Some ecologists believe that ecosystem stability is largely the result of species diversity. The higher the diversity, they say, the greater the stability. In support of this idea are observations of extremely complex ecosystems, such as rain forests, remain unchanged almost indefinitely if undisturbed. Simple ecosystems such as the tundra are more volatile. They can experience sudden, drastic shifts in population size. Other simplified ecosystems such as fields of wheat and corn also show extreme vulnerability to change, and they collapse if abiotic or biotic factors shift.

To see how ecologists explain this phenomenon, look at the differences in food webs in simple and complex ecosystems. As illustrated in Figure 2.15, the number of species in a food web in a mature ecosystem is large. So is the number of interactions among these organisms. In a complex ecosystem the elimination of one species would probably have little effect on the ecosystem balance. In sharp contrast, the number of species in the food web of a simple ecosystem is small. The elimination of one species could have repercussions on all other species.

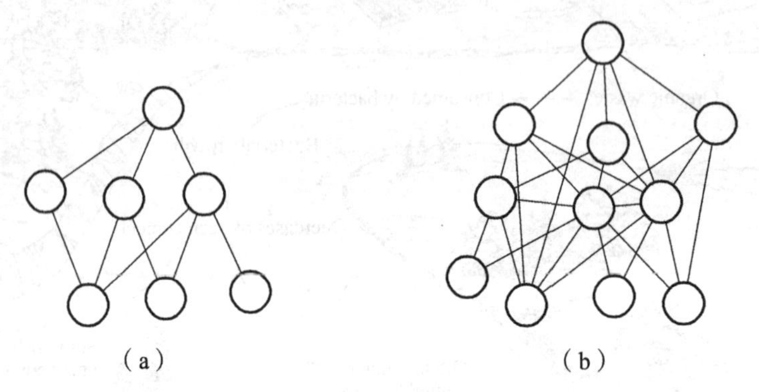

Figure 2.15 (a) Food web in a simple ecosystem. Circles represent organisms. (b) Food web in a complex ecosystem

2.3.3 Human Impact on Ecosystems

Throughout human history, we have used our knowledge to gain control over the environment – to shape it to our liking and enhance our survival. But our understandable search for control and security has not always produced the desired outcome. Water hyacinth-choked rivers, polluted streams, acidic rainfall, radioactive spills, energy shortages, runaway population growth, and a host of other common maladies are often the unanticipated results of our actions.

This section examines two general ways in which humans alter the biosphere: they tamper with either the biotic or the abiotic factors. What we do to the biosphere, we also do to ourselves.

Tampering with Biotic Factors

Introducing Competitors The killer bee is actually an African honeybee which lives in hives with 30,000 to 80,000 ill-tempered cohorts. It was brought to the New World quite intentionally. In 1956 a geneticist, Warwick Kerr, imported some African honeybees to Venezuela in an attempt to develop a successful stock of honey producers. The docile European bees used previously had fared poorly in the tropical climate. Kerr thought that interbreeding the two might yield a more successful tropical strain.

Knowing their aggressiveness, the researcher isolated the bees in screened-in hives. But in 1957, a visitor unwittingly lifted the screen, allowing 26 queens to escape with their entourage. Trouble soon began. The killer bees quickly spread, moving a remarkable 350 to 500 kilometers a year, interbreeding with honeybees, and destroying the honey industry in Venezuela and other countries. And that's not all. Bees assaulted people, horses, and livestock that crossed their paths.

Killer bees arrived in the United States by ship, many experts think, accompanying a load of pipe from South America. So far, California officials have located at least four colonies and have destroyed hundreds of commercial beehives to destroy any killer bees that may have mixed with the colonies.

Experts worry about the impact of the killer bees on the US honey industry. Should it spread into northern climates and breed with its tamer cousin, the killer bee could put an end to honey production. Hives would have to be destroyed to prevent further spread and attacks. The loss of honey production would, however, be minor compared with the indirect effect of losing America's honeybees. Each year honeybees pollinate 90 major crops.

Many biologists hoped that the northward spread of the killer bee would be halted by the colder climates. A study showed that killer bees could survive at a 0 °C temperature for six months. It is feared that the bees could migrate as far north as Canada, causing widespread damage to the North American honey industry and to crops pollinated by native bees.

The saga of the African honeybee is one of a number of biological nightmares created by the introduction of a foreign species into a new region. But not all such introductions have adverse effects. The ring-necked pheasant and chukar partridge, both aliens in this country, have done well in some areas. In other cases alien species have perished without a trace. Hardy species such as the killer bee, however, are the ones that demand our attention and remind us of the folly of careless introductions.

Consider what happened when rabbits were intentionally introduced into Australia. Twelve pairs of European rabbits were brought into the country in 1859 and released on a private ranch. In a few years the rabbits had proliferated wildly and begun eating grass intended for sheep, creating a major national environmental problem. Five rabbits eat as much grass as one sheep. Despite a major campaign to remove rabbits from Australia, by 1953 over a billion were inhabiting 3 million square kilometers (1.2 million square miles).

Plants such as the prickly pear cactus or water hyacinth can also reproduce uncontrollably in foreign environments. Taking over a new territory, they can wipe out native populations that compete for the same habitat. The results can be ecologically and economically disastrous.

Eliminating or Introducing Predators Predators have never fared well in human societies. Early hunters and gatherers killed them for food because they viewed them as competition for prey. Modern societies have carried on this dangerous tradition, killing bears, eagles, hawks, wolves, coyotes, and mountain lions with a vengeance, which often leads to serious ecological consequences.

One such example took place on the Kaibab Plateau, on the north rim of the Grand Canyon. In the early 1900s, the state of Arizona put a bounty on wolves, coyotes, and mountain lions, which triggered their wholesale slaughter. Within 15 years virtually all of these predators had been eliminated. This intervention spawned an ecological catastrophe. Without its natural control, the deer population soared from 4,000 to about 100,000 by 1924. The deer overgrazed the plateau; approximately 60,000 died from starvation in the following winters. The vegetation has still not completely recovered yet.

Occasionally, however, predators are introduced into areas with adverse effects. The mosquito fish, a native of the southeastern United States, has been introduced into many subtropical regions throughout the world because it eats the larvae of mosquitoes and therefore helps to control malaria, a mosquito-borne disease. Unfortunately, the mosquito fish also feeds heavily on zooplankton, single-celled organisms that consume algae. By depleting zooplankton populations, the mosquito fish removes environmental resistance that curbs algal growth. This causes algae to proliferate and form thick mats that reduce light penetration and plant growth in aquatic ecosystems.

These examples illustrate that altering the trophic structure by introducing or eliminating predators can drastically affect ecosystems and human populations as well.

Introducing Disease Organisms Pathogenic (disease-producing) organisms are a natural part of ecosystems. Humans have unwittingly introduced pathogens into new environments where there are no natural controls. There, they have reproduced at a high rate and caused serious damage.

In the late 1800s, for example, a fungus that infected Chinese chestnuts was accidentally introduced into the United States. It had been carried in with several Chinese chestnut trees brought to the New York Zoological Park. The Chinese chestnut soon evolved mechanisms to combat the fungus and was immune to it. But the American chestnut had no resistance at all and was virtually eliminated from this country between 1910 and 1940.

Tampering with Abiotic Factors

Humans also tamper with many abiotic factors by polluting the air, water, and land, and by depleting resources, such as water. The consequences can be as significant as those from our manipulation of the biotic world.

Pollution Water pollution and air pollution create an unfavorable environment for many living organisms. Chlorinated compounds released into rivers from wastewater treatment plants, for example, can virtually eliminate native fish. Oil spills on lakes, rivers, and oceans destroy fish, reptiles, and birds. Toxic pesticides eliminate birds that feed on contaminated insects and fish. Thermal pollution from power plants kills fish and many aquatic organisms on which fish feed. Possible global changes in temperature brought about by increasing atmospheric carbon dioxide could alter climate in many regions of the world, possibly eliminating thousands of species of plants and animals. In these and many other cases, human activities create an unfavorable abiotic environment that can reduce or eliminate species and upset the ecological balance.

Resource Depletion Human populations deplete or destroy resources used by other species, too. The diversion of mountain streams to supply growing cities, for instance, has left many waterways dry. Housing developments along coastlines are often made possible by filling in estuaries and marshes with dirt.

Simplifying Ecosystems

Tampering with abiotic and biotic factors tends to simplify an ecosystem by reducing species diversity. A reduction in species diversity may cause imbalance and eventual collapse of an ecosystem.

Ecosystem simplification is best seen when natural ecosystems are converted into farmland. Grassland contains many species of plants and animals. When plowed under and planted in one crop, called a monoculture, the field becomes simplified and vulnerable to insects, disease, drought, wind, and adverse weather.

The reasons for this susceptibility are many. Perhaps one of the most important is that monocultures provide a virtually unlimited food source for insects and plant pathogens, especially viruses and fungi. As crops grow, food supplies increase dramatically, favoring massive growth of pest populations. Viruses, fungi, and insects become major pests. Monocultures provide little or no environmental resistance.

The need to protect monocultures leads to many far-reaching problems. For instance, farmers have long used chemical pesticides to control outbreaks of pests. Chemical pesticides used to control fungi, viruses, and insects may be carried in the air or water to natural ecosystems, where they may poison beneficial organisms such as honeybees. The widespread use of DDT contaminated many terrestrial and aquatic ecosystems. Passed through the food chain, DDT and its chief breakdown product DDE reached high levels in top-level consumers, including ospreys, brown pelicans, and peregrine falcons.

This poison did not kill the birds outright; rather, it interfered with the deposition of calcium in their eggshells, resulting in thinning of the shells. The fragile eggs were easily broken. Few embryos survived, and populations fell sharply, reminding us of an important biological principle: an organism lives as it breeds.

The DDT incident also illustrates an important biological phenomenon called biological magnification or simply biomagnification, the accumulation of certain substances in food chains, with the highest concentrations found in the highest trophic levels. Because of this phenomenon, fairly low levels of DDT in the environment can result in dangerously high levels in organisms.

The peregrine falcon, which nests on rocky ledges throughout the United States, was nearly destroyed by DDT. By the time scientists had determined that the decline in reproduction was the result of DDT and DDE.

The story of the peregrine falcon, like so many other species that have been brought back from the brink of extinction, is one of personal dedication and triumph. By using ecological caution in releasing a powerful toxin into the environment, however, we could have avoided a great deal of the cost and effort.

2.3.4 Reestablishing the Balance

With industrialization, the damage increased dramatically. During that time, people cut the trees down to build homes and factories. Dam-builders cut off the flow of fresh water to coastal freshwater swamps, letting in the sea and destroying the habitat for fish and other animals. Some of these devastated lands are now on the mend.

The Birth of a New Science In 1985, a man by the name of Ed Garbisch arrived in New Jersey to begin the long and costly process of restoring an ancient freshwater wetland. Garbisch and others like him gave rise to a new field of ecology, called restoration ecology. Scientists often prefer to call this new science conservation biology, the study of how ecosystem recovery occurs and how it can be facilitated. In this relatively new field of study, the principles of science are used to repair land and waters damaged by decades of misuse. In many respects, conservation biology is the rehabilitative medicine of the ecological sciences. Some of the earliest restoration projects were designed to reclaim badly damaged land that had been surface-mined for coal and other minerals. Today, however, ecologists and others are working to restore marshes, tropical rain forests, streams, and prairies throughout the world.

John Berger, who held a PhD in ecology, has done as much to stimulate restoration as any person. His book *Restoring the Earth* describes what people are doing in the United States to correct past mistakes. Berger also founded a nonprofit environmental group called Restoring the Earth, whose purpose is to repair the damaged American landscape. Because of-their work and the work of John Berget and Ed Garbisch, hundreds of marshes have been restored along the east coast from Maine to Virginia, reestablishing native plants and habitats for many species.

Benefits of Restoration Restoring marshes has economic as well as aesthetic benefits. To protect an eroding shoreline with concrete or rocks, for example, cost much more than that of planting a 7-meter-wide strip of salt marsh. Plants can also turn a barren, desolate beachfront into dense, lush, and emerald marshes.

A marsh's value cannot be easily calculated in dollars and cents. Biologists in Louisiana estimated some years ago that an acre of salt marsh was worth $82,000 solely for its ability to reduce pollutants such as sediment.

Marshes are one of the world's most productive ecosystems. A single hectare of marsh produces more than 25 tons of organic matter per year, which is more than twice the yield of a corn field and 10 times the yield of coastal waters.

Many people see marshes as muddy, uninhabitable places that should be drained or filled to build houses, airports, industrial parks, or garbage dumps. Today, approximately one half of the salt marshes have already been destroyed. When a salt marsh disappears, shorelines erode at an accelerated rate, fish populations collapse, birds vanish, wildlife retreats, and some of nature's remarkable plant communities are destroyed.

Chapter 3
Water Pollution

Water is one of the material bases for human survival, and the main component of all organisms. It plays an important role in regulating global climate, transporting materials etc.. As the world's population and the industrial products increase, water consumption will also accelerate, accordingly more and more produced waste water is. Once the waste water is discharged into our waterways with little or no treatment, it will cause water pollution, which will result in the destruction and crisis of water resources.

Water pollution intensified following World War II when dramatic increases in urban density and industrialization occurred. In developed countries, the beaches are polluted with medical wastes and leftover sludge from wastewater treatment plants. Our traditional confidence in the quality of our drinking water has been seriously shaken as we find potential carcinogens in our groundwater and as we come to realize that we have been creating carcinogens by chlorinating our surface water. In less developed regions, wastes from burgeoning populations are a threat to public health and endanger the continuing use of the scarce water supplies

This chapter will deal primarily with the water pollutants and the measurement of water quality. Besides, the effluent standards and water quality requirements are introduced, and the sources of water pollution presented. The principles of wastewater treatment and the municipal wastewater treatment system are explained.

3.1 WATER AND WATER POLLUTION

3.1.1 Concept of Water Pollution

Water pollution refers to degradation of water quality as measured by biological, chemical or physical criteria. Degradation of water is generally judged in terms of the intended use of the water, departure from norm, effects on public health, or ecologic impacts.Water pollution is global in scope, but the type of pollution varies according to a country's level of development.

In the less developed nations water pollution is predominantly caused by human and animal wastes, pathogenic organisms from this waste, and sediment from unsound farming and timbering practices. The developed nations also suffer from these problems, but with their extravagant life-styles and widespread industry they create an additional assortment of potentially hazardous pollutants: heat, toxic metals, acids, pesticides, and organic chemicals.

From a public health or ecologic point of view, a pollutant is any biological, physical, or chemical substance in which an identifiable excess is known to be harmful to other desirable living organisms. Like air pollutants, water pollutants come from numerous natural and anthropogenic sources. Because water respects no boundaries, pollutants produced in one country, often end up in another's drinking or bathing water. The thoughtless dumping of wastes in rivers, unfortunate accidents, and uncontrolled growth of pollution can have dire consequences for commercial fisheries.

Movement of pollutants from lakes and rivers to oceans is only half the problem. In recent years scientists have revealed cross-media contamination, that is, the movement of a pollutant from one medium (air) to another (water). Pesticides sprayed on corps can drift to nearby lakes and, from there, flow to the oceans. Toxic organics dumped in evaporating ponds ascend to the clouds, only to rain down on land and lakes. Hazardous wastes buried in the ground leak into aquifers, whose water replenish streams.

3.1.2 Water Pollutants

According to the characteristics of polluted materials, water pollutants can be divided into the following three categories: chemical pollutants, physical pollutants, and biological pollutants.

Chemical Pollutants

Non-toxic inorganic substances Non-toxic inorganic substances include acid, alkali and some inorganic salts, which are discharged by many industries, such as dye industry, paper-making industry, alkali-making industry, etc.. They make water pH value change, improve the hardness of water and reduce dissolved oxygen in water.

Toxic inorganic substances Toxic inorganic substances are divided into two categories: metals and non-metals. The toxic metals mainly refer to heavy metals whose specific gravities are greater than 4-5 g/cm^3. The heavy metals in wastewater are mainly mercury, chromium, cadmium, lead, zinc, nickel, copper, cobalt, manganese, titanium, vanadium, molybdenum, antimony and bismuth, etc.. Toxic metals cannot be biodegraded. Some of them may be transformed to more toxic substances which can be concentrated in organisms and endanger life-forms and humans through the food chain. The Minamata incident in Japan in 1953 is an example. Methyl mercury was concentrated in the fish through the food chain and threatened people's life. The toxic non-metals in wastewater are mainly arsenic, selenium, fluorine, sulfur, cyanide and nitrite ion, etc..

Toxic organic substances The common toxic organic substances are phenolic compounds, organic pesticides, polycyclic aromatic hydrocarbons (PAH), polychlorinated biphenyls (PCB), detergent and aromatic amines, etc. Many of them are artificially synthesized and difficult to be biodegraded due to their stable chemical properties. Some even have carcinogenic or mutagenic effect.

Oxygen-demanding material Anything that can be oxidized in the receiving water with the consumption of dissolved molecular oxygen is termed ***oxygen-demanding material***. This material is usually biodegradable organic matter but also includes certain inorganic compounds. The consumption of dissolved oxygen (DO) poses a threat to higher forms of aquatic life which must have oxygen to live. The critical level of DO varies greatly among species. For example, brook trout may require about 7.5 mg/L of DO while carp may survive at 3 mg/L. As a rule, the most desirable commercial and game fish require high levels of dissolved oxygen. Oxygen-demanding materials in domestic sewage come primarily from human wastes and food residue. Particularly noteworthy among the many industries which produce oxygen-demanding wastes are the food processors and the paper industry. Almost any organic matter emerging naturally, such as animal droppings, crop residues, or leaves, which get into the water from nonpoint sources, contribute to the depletion of DO.

Nutrients Nitrogen and phosphorus, the two nutrients of primary concern, are considered pollutants because they are too much of a good thing. All living things require these nutrients for growth. Thus, they must be present in rivers and lakes to support the natural food chain. Problems arise when nutrient levels become excessive and the food web is grossly disturbed, which causes some organisms to proliferate at the expense of others. Excessive nutrients often lead to large growths of algae, which in turn become oxygen-demanding material when they die and settle to the bottom. Some major sources of nutrients are phosphorus-based detergents, fertilizers, and food-processing wastes.

Oil Oil, discharged into surface water (usually the ocean), are mainly derived from petroleum transport, submarine oil drilling operations, oil tanker accident, industrial wastewater emissions and tanker washing etc.. It can form a layer of oil film on the surface of the water, which isolates the water from the air and causes water hypoxia. Oil spills from submarine oil drilling operations may cause damage to beaches and marine life if the oil drifted ashore. Polycyclic aromatic hydrocarbon , which has carcinogenesis, is contained in oil and can enter the body through the food chain.

Physical Pollutants

Suspended solids Organic or inorganic particles which are carried by the wastewater into a receiving water are termed ***suspended solids*** (SS). When flowing into a pool or a lake, the speed of the water is reduced and many of these particles settle to the bottom as sediment. In common usage, the term sediment also includes eroded soil particles which are being carried by water even if they have not yet settled. Colloidal particles which do not settle readily cause the turbidity found in many surface waters. Organic suspended solids may also exert an oxygen demand. Inorganic suspended solids are discharged by some industries but result mostly from soil erosion which is particularly bad in areas of logging, strip mining and construction activity. As excessive sediment loads are deposited into lakes and reservoirs, their usefulness is reduced. Even in rapidly moving mountain streams, sediment from mining and logging operations has destroyed many living places for aquatic organisms. For example, salmon eggs can only develop and hatch in loose gravel stream beds. As the pores between the pebbles are filled with sediment, the eggs suffocate and the salmon population is reduced.

Heat Although heat is not often recognized as a pollutant, those in the electric power industry are well aware of the problems of disposing of waste heat. Also, many industrial process waters are much hotter than the receiving waters. In some environments an increase of water temperature can be beneficial. For example, production of clams and oysters can be increased in some areas by warming the water. On the other hand, increases in water temperature can have negative impacts. Many important commercial and game fish such as salmon and trout, can only live in cool water. In some instances the discharge of heated water from a power plant can completely block salmon migration. Higher temperatures also increase the rate of oxygen depletion in areas where oxygen-demanding wastes are present.

Radioactive Materials Radioactive materials in water may be dangerous pollutants. Impurities containing radioactive elements form a special kind of pollution sources, which are known as the radioactive pollution. Of particular concern are possible effects on people, other animals and plants of long-term exposure to low doses of radioactivity.

Biological Pollutants

Pathogenic organisms Microorganisms found in wastewater include bacteria, viruses, and protozoa which come from domestic sewage, especially the hospital sewage and some industrial wastewater, such as biological products, tanning, brewing, slaughter wastewater etc.. When discharged into surface waters, they make the water unfit for drinking. If the concentration of pathogens is sufficiently high, the water may also be unsafe for swimming and fishing. Certain shellfish can be toxic because they concentrate pathogenic organisms in their tissues, making the toxicity levels in the shellfish much greater than the levels in the surrounding water.

3.2 MEASUREMENT OF WATER QUALITY

Quantitative measurements of pollutants are obviously necessary before water pollution can be controlled. The measurement of these pollutants is, however, fraught with difficulties.

The first problem is that the specific materials responsible for the pollution are sometimes not known. The second difficulty is that these pollutants are generally at low concentrations, and very accurate methods of detection are therefore required.

Many of the pollutants are measured in terms of milligrams of the substance per liter of water (mg/L). This is a weight/volume measurement. In many older publications, pollutants are measured as parts per million (ppm), a weight/weight parameter. If the liquid involved is water, these two units are identical, since 1 milliliter of water weighs 1 gram. Because of the possibility of some wastes not having the specific gravity of water, the ppm measure has been scrapped in favor of mg/L.

3.2.1 Dissolved Oxygen

Probably the most important measure of water quality is the DO. Oxygen, although poorly soluble in water, is fundamental to aquatic life. Without free DO, streams and lakes become uninhabitable to gill-breathing aquatic organisms. Dissolved oxygen is inversely proportional to temperature. The saturation value decreases rapidly with increasing water temperature, as shown in Table 3.1. The balance between saturation and depletion is therefore tenuous.

Table 3.1 Solubility of oxygen

Temperature of Water (°C)	Saturation Concentration of Oxygen in Water (ml/L)
0	14.6
2	13.8
4	13.1
6	12.5
8	11.9
10	11.3
12	10.8
14	10.4
16	10.0
18	9.5
20	9.2
22	8.8
24	8.5
26	8.2
28	8.0
30	7.6

The amount of oxygen dissolved in water is usually measured either with an oxygen probe or by iodometric titration: Winkler DO Test. The Winkler test for DO, developed almost 100 years ago, is the standard to which all other methods are compared.

3.2.2 Biochemical Oxygen Demand

When biodegradable organic matter is released into a body of water, microorganisms, especially bacteria, feed on the wastes, breaking it down into simpler organic and inorganic substances. When this decomposition takes place in an aerobic environment, that is, in the presence of oxygen, it will produce nonobjectionable, stable end products such as carbon dioxide (CO_2), sulfate (SO_4^{2-}), orthophosphate (PO_4^{3-}), and nitrate (NO_3^-). A simplified representation of aerobic decomposition is given by the following:

$$\text{Organic matter}+O_2 \xrightarrow{\text{microorganisms}} CO_2+H_2O+\text{New cells}+\text{Stable products}$$

When insufficient oxygen is available, the resulting anaerobic decomposition is performed by completely different microorganisms. They produce end products that can be highly objectionable, including hydrogen sulfide (H_2S), ammonia (NH_3), and methane (CH_4). Anaerobic decomposition can be represented by the following:

$$\text{Organic matter} \xrightarrow{\text{microorganisms}} CO_2+CH_4+\text{New cells}+\text{Unstable products}$$

The methane produced is physically stable, biologically degradable, and a potent greenhouse gas. When emitted from bodies of water it is often called swamp gas. It is also generated in the anaerobic environment of landfills, where it is sometimes collected and used as an energy source.

The amount of oxygen required by microorganisms to oxidize organic wastes aerobically is called the ***biochemical oxygen demand*** (BOD). BOD may have various units, but most often it is expressed in grams of oxygen required per liter of wastes (g/L) or the equivalent kg/m^3.

Five-Day BOD Test

The biological oxidation process of organic matter is a slow process (Figure 3.2). The experimental data show that, for the majority of organic matter, the amount of oxygen that will be required for biodegradation for 20 days is about 95%-99%; for 5 days about 70%-80%. The total amount of the oxygen is an important measure of the impact that a given waste stream will have on the receiving body of water. While we could imagine a test in which the oxygen required to completely degrade a sample of waste would be measured, such a test would require an extended period of time (several weeks), making it impractical. As a result, it has become standard practice simply to measure and report the oxygen demand over a shorter, restricted period of 5 days, realizing that the ultimate demand is considerably higher.

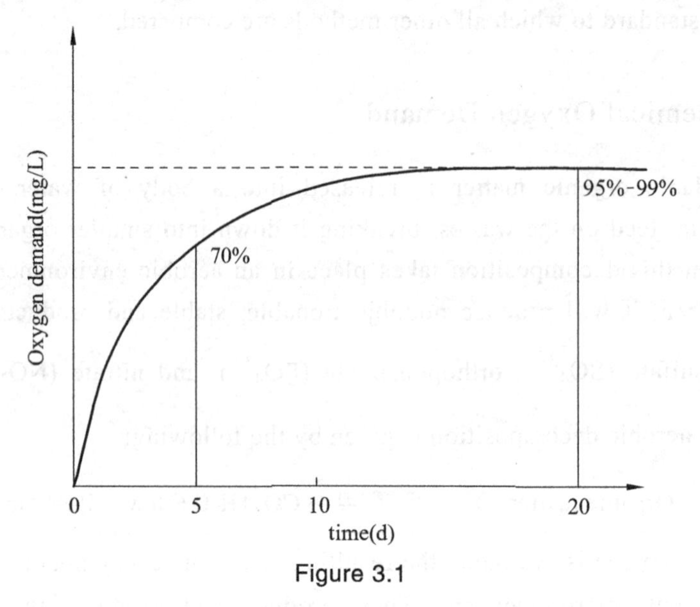

Figure 3.1

The 5-day BOD, or **BOD_5**, is the total amount of oxygen consumed by microorganisms during the first 5 days of biodegradation. In its simplest form, a BOD_5 test would involve putting a sample of waste into a stoppered bottle, measuring the concentration of dissolved oxygen in the sample at the beginning of the test and repeat the process again 5 days later. The difference in DO would be the 5-day BOD. Light must be kept out of the bottle to keep algae from adding oxygen by photosynthesis and the stopper is used to keep air from replenishing DO that has been removed by biodegradation. To standardize the procedure, the test is run at a fixed temperature of 20 °C. Since the oxygen demand of typical waste is several hundred milligrams per liter, and since the saturated value of DO for water at 20 °C is only 9.1 mg/L, it is usually necessary to dilute the sample to keep final DO above zero. If during the 5 days the DO drops to zero, then the test is invalid, since more oxygen would have been removed if been available.

The 5-day BOD of a diluted sample is given by

$$BOD_5 = \frac{DO_i - DO_f}{P} \tag{3-1}$$

where DO_i = the initial dissolved oxygen of the diluted wastewater

DO_f = the final dissolved oxygen of the diluted wastewater

P = the dilution fraction= $\frac{\text{Volume of wastewater}}{\text{Volume of wastewater plus dilution water}}$

A standard BOD bottle holds 300 mL, so P is just the volume of wastewater divided by 300 mL.

Example 3.1 Unseeded 5-Day BOD Test

A 10.0-mL sample of sewage mixed with enough water to fill a 300-mL bottle has an initial DO of 9.0 mg/L. To help assure an accurate test, it is desirable to have at least a 2.0 mg/L drop in DO during the 5-day run, and the final DO should be at least 2.0 mg/L. For what range of BOD_5 would this dilution produce the desired results?

Solution The dilution fraction is P=10/300. To get at least a 2.0 mg/L drop in DO, using (3-1)

$$BOD_5 \geqslant \frac{DO_i - DO_f}{P} = \frac{2.0 \text{ mg/L}}{10/300} = 60 \text{ mg/L}$$

and to assure at least 2.0 mg/L of DO remaining after 5 days requires that

$$BOD_5 \leqslant \frac{9.0 - 2.0 \text{ mg/L}}{10/300} = 210 \text{ mg/L}$$

So this dilution will be satisfactory for BOD_5 values between 60 and 210 mg/L.

So far, we have assumed that the dilution water added to the waste sample has no BOD of its own, which would be the case if pure water were added. In some cases it is necessary to seed the dilution water with microorganisms to assure that there is an adequate bacterial population to carry out the biodegradation. In such cases, to find the BOD of the waste itself, it is necessary to subtract the oxygen demand caused by the seed from the demand in the mixed sample of waste and dilution water.

3.2.3 Chemical Oxygen Demand

The amount of oxygen required by a strong chemical oxidizing agent to oxidize wastes in an acid medium is called the ***chemical oxygen demand*** (COD). Here, the wastes mainly refer to the organics, meanwhile, they also include reducibility inorganics. Because nearly all organics and reducibility inorganics are oxidized in the COD test and only some are decomposed during the BOD test, the COD values are always higher than the BOD values of the same wastewater. The difference between them can roughly show the amount of wastes that cannot be biodegraded.

Potassium dichromate is generally used as an oxidizing agent. It is an inexpensive compound that is available in very pure form. A known amount of this compound is added to a measured amount of sample, and the mixture is boiled. The reaction, in unbalanced form, is

$$C_xH_yO_z + Cr_2O_7^- + H^+ \xrightarrow{\Delta} CO_2 + H_2O + Cr^{3+}$$
(organic) (dichromate)

After boiling with an acid, the excess dichromate (not used for oxidizing) is measured by adding a reducing agent, usually ferrous ammonium sulfate. The difference between the chromate originally added and the chromate remaining is the chromate used for oxidizing the organics. The more chromate used, the more organics were in the sample, and hence the higher the COD.

Potassium permanganate is another kind of oxidizing agent. In order to distinguish them, we usually use COD_{Mn} to express potassium permanganate chemical oxygen demand, while COD_{Cr} for potassium dichromate chemical oxygen demand.

The BOD test takes more time than COD test which takes only a matter of hours. Some organic materials, such as cellulose, phenols, benzene, and tannic acid, resist biodegradation. Others, such as pesticides and various industrial chemicals, are nonbiodegradable because they are toxic to microorganisms. The chemical oxygen demand, COD, is a measured quantity that depends neither on the ability of microorganisms to degrade the waste nor on knowledge of the particular substances in question. In a COD test, a strong chemical oxidizing agent is used to oxidize the organics rather than relying on microorganisms to do the job. However, it does not distinguish between the oxygen demand that will actually be felt in a natural environment due to biodegradation, and the chemical oxidation of inert organic matter. Nor does it provide any information on the rate at which actual biodegradation will take place. The measured value of COD is higher than BOD, though for easily biodegradable matter the value of the two will be quite similar.

COD and BOD are two important indexes of wastewater. They are independent, but complementary. The COD test is sometimes used as a way to estimate the ultimate BOD. The ratio of BOD_5 and COD can be used as an important index for selecting biological treatment of organic wastewater or not. The higher the ratio is, the more the contents of biodegradable organic matter are, and hence the better the biological treatment effect will be. In general, if the ratio is greater than or equal to 0.3, the wastewater can be treated in a biological way, if not, it cannot be biodegraded and should be treated by other methods.

3.2.4 Total Organic Carbon

Total organic carbon (TOC) is the total carbon content of organic pollutants in wastewater. Since the ultimate oxidation of organic carbon is to CO_2, the total combustion of a sample will yield some significant information on the amount of organic carbon present in a wastewater sample. Without elaboration, this is done by allowing a small portion of the sample to be burned in a combustion tube and measuring the amount of CO_2 emitted. This test is not widely used at present, mainly because of the expensive instrumentation required.

3.2.5 pH

The pH of a solution is a measure of hydrogen ion concentration. An abundance of hydrogen ions makes a solution acid, whereas a dearth of H^+ ions makes it basic, A basic solution has, instead, an abundance of hydroxide ions, OH^-.

pH value is the negative logarithm of $c(H^+)$, so that

$$\mathrm{pH} = -\log[\mathrm{H}^+] = \log\frac{1}{[\mathrm{H}^+]}$$

Where, $[H^+]$ is the concentration of H^+ ions, the unit is mol/L.

In any aqueous solution, the product of H^+ and OH^- concentrations is always constant, or the ion-product constant of water(K_W) which is only related with temperature. When the temperature is 25°C, the value of K_W is 1×10^{-14}, or

$$[\mathrm{H}^+]\times[\mathrm{OH}^-] = K_W = 1\times10^{-14}\text{, or}$$

$$\mathrm{pH} + \mathrm{pOH} = \mathrm{p}K_W = 14$$

where, K_W is the ion-product constant of water; $\mathrm{pOH} = -\lg[\mathrm{OH}^-]$; $\mathrm{p}K_W = -\lg[K_W]$.

When $[H^+] = [OH^-]$, pH=pOH, so that $\mathrm{pH} = \frac{1}{2}\mathrm{p}K_W = 7$. The solution is neutral (neither acidic nor basic); When $[H^+] > [OH^-]$, $pH < 7$, the solution is acidic. On the contrary, When $[H^+] < [OH^-]$, $pH > 7$, the solution is alkaline . The pH range of dilute solutions is from 0 to 14.

Commonly, the pH value of domestic sewage is between 7.2-7.6, while the pH value of industrial wastewater varies violently.

The measurement of pH is now almost universally by electronic means. Electrodes that are sensitive to hydrogen ion concentration (strictly speaking, the hydrogen ion activity) convert the signal to electrical current.

pH is important in almost all phases of water and wastewater treatment. Aquatic organisms are sensitive to pH changes, and biological treatment requires either pH control or monitoring. In water treatment pH is important in ensuring proper chemical treatment as well as in disinfection and corrosion control. Mine drainage often involves the formation of sulfuric acid, which is extremely detrimental to aquatic life. Continuous acid deposition from the atmosphere may lower the pH of a lake substantially.

3.2.6 Solids

One of the main problems with wastewater treatment is that so much of the wastewater is actually solids. The separation of these solids from the water is in fact one of the primary objectives of wastewater treatment.

Strictly speaking, in wastewater anything other than water would be classified as a solid. The usual definition of solids, however, is the residue on evaporation at 103 °C. The solids thus measured are known as total solids. Total solids may be divided into two categories: the dissolved solids and the suspended solids. If we put a teaspoonful of common table salt in a glass of water, the salt will dissolve. The water will not look any different, but the salt will remain behind if we evaporate the water. A spoonful of sand, however, will not dissolve and will remain as sand grains in the water. The salt is an example of a dissolved solid, whereas the sand would be measured as a suspended solid.

The separation of suspended solids from dissolved solids is by means of a special crucible, called a Gooch crucible. As shown in Figure 3.2, the Gooch crucible has holes on the bottom on which a glass fiber filter is placed. The sample is then drawn through the crucible with the aid of a vacuum. The suspended material is retained on the filter, while the dissolved one passes through. If the initial dry weight of the crucible and filter are known, the subtraction of this from the total weight of crucible, filter, and the dried solids caught on the filter yields the weight of suspended solids, expressed as milligrams per liter.

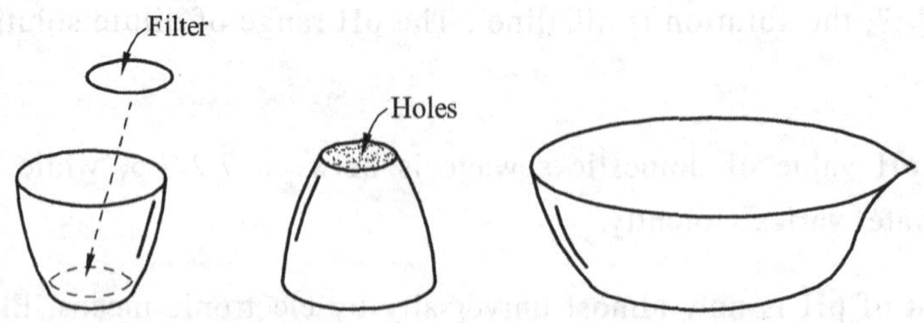

Figure 3.2 The gooch crucible for suspended solids (with filter) and the evaporating dish for total solids

(Source: P. Aarne Vesilind, *Environmental Pollution And Control*, Ann Arbor Science, 1975, P41)

Solids may be classified in another way: those that are volatilized at a high temperature and those that are not. The former are known as volatile solids, the latter as fixed solids. Usually, volatile solids are organic. Obviously, at 600 °C, the temperature at which the combustion takes place, some of the inorganics are decomposed and volatilized.

The relationship between the total solids and the total volatile solids can best be illustrated by an example.

Example 3.2

Given the following data:

- weight of dish (such as shown in Figure 3.2)=48.621 g
- 100 mL of sample is placed in dish and evaporated. Weight of dish and dry solids= 48.6432 g
- Dish is placed in 600 °C furnace, then cooled. Weight=48.6300 g.

Find the total, the volatile, and the fixed solids.

$$\begin{aligned}\text{Total Solids} &= \frac{(\text{dish+dry solids})-(\text{dish})}{\text{volume of sample}} \\ &= \frac{48.6432-48.6212}{100} \\ &= 220\times10^{-6}\ \text{g/mL} \\ &= 220\times10^{-3}\ \text{mg/mL} \\ &= 220\ \text{mg/L}\end{aligned}$$

$$\begin{aligned}\text{Fixed Solids} &= \frac{(\text{dish+unburned solids})-(\text{dish})}{\text{volume of sample}} \\ &= \frac{48.6300-48.6212}{100} \\ &= 88\ \text{mg/L}\end{aligned}$$

$$\begin{aligned}\text{Volatile Solids} &= \text{Total Solids} - \text{Total Fixed Solids} \\ &= 220-88=132\ \text{mg/L}\end{aligned}$$

It is often necessary to measure the volatile fraction of suspended material, since this is a quick measure of the amount of microorganisms present. The volatile suspended solids are determined by simply burning the Gooch crucible and weighing it again. The loss in weight is interpreted as volatile suspended solids.

3.2.7 Bacteriological Measurements

From the public health standpoint, the bacteriological quality of water is as important as the chemical quality. A number of diseases may be transmitted by water, among which are typhoid and cholera. There are many pathogens in wastewater. Each has a specific detection procedure and must be screened individually. The concentration of these organisms may be too small to be detected, just like a needle in a haystack. And yet only one or two organisms in the water might be sufficient to cause an infection.

How then can we measure for bacteriological quality? The answer lies in the concept of indicator organisms. The indicator most often used is a group of microbes of the family coliform bacteria, which are organisms normal to the digestive tracts of warm-blooded animals. In addition to that attribute, coliforms are plentiful and easily detected with a simple test. Generally, coliforms are harmless except in unusual circumstances and can hardily survive longer than most known pathogens.

Coliforms have thus become universal indicator organisms. But the presence of coliforms does not prove the presence of pathogens. If a large number of coliforms are present, there is a good chance of recent pollution by wastes from warm-blooded animals, and therefore the water may contain pathogenic organisms.

The last point should be emphasized. The presence of coliforms does not mean that there are pathogens in the water. It simply means that there might be. A high coliform count is thus suspicious, and the water should not be consumed (although it may be perfectly safe).

3.2.8 Viruses

Because of their minute size and extremely low concentration and the need to culture them on living tissues, pathogenic viruses are fiendishly difficult to measure. Because of this problem, there are as yet no standards for viral quality of water supplies.

One possible method of overcoming this difficulty is to use an indicator organism, much like the coliform group is used as an indicator for bacterial contamination. This can be done by using a bacteriophage, or a virus that attacks only a certain type of bacterium. For example, coliphages attack coliform organisms and, because of their association with wastes from warm-blooded animals, seem to be ideal indicators. The test for coliphages is performed by inoculating a Petri dish containing an ample supply of a specific type of coliform with the wastewater sample. Coliphages will attack the coliforms, leaving visible can be made spots or plaques that can be counted, and an estimate of the number of coliphages per known volume can be made.

3.3 EFFLUENT STANDARDS AND WATER QUALITY REQUIREMENTS

3.3.1 Effluent Standards

Effluent standards, rigid limits enforced by law, are important parts of the environmental protection standards system. The standards of the United States have always been the reference standards to other countries for their standards setting. This part will focus on introducing American effluent standards.

Under the *Clean Water Act* (CWA) the Environmental Protection Agency (EPA) oversees and states operate programs designed to reduce the flow of pollutants into natural watercourses. Every municipal or industrial treatment facility that discharges wastewater effluent into the environment must obtain a National Pollutant Discharge Elimination System (NPDES) permit. The NPDES permit clearly states the allowable amounts of pollutants that a particular facility can discharge, and specifies a compliance schedule if a discharge cannot immediately meet the required limitations. (Businesses discharging to a sewer system rather than a natural watercourse are not required to obtain an NPDES permit; however, they must obtain permits from the municipal treatment plants receiving the waste.) While detractors have labeled these "permits to continue pollution," the permitting system has, nevertheless, had a major beneficial effect on the quality of surface water. Typical effluent standards for a domestic wastewater treatment plant in the United States range from 5 to 20 mg/L, for example. The intent is to tighten these limits as required to enhance water quality.

The efforts to improve water quality also have focused on establishing total maximum daily loads (TMDLs) for various pollutants in watersheds. In addition, nonpoint sources of pollution are being addressed. In particular, regulations concerning stormwater and combined sewer overflows (CSOs) are being implemented.

Treatment Efficiency Treatment efficiency can be defined as the ratio of the amount of pollutants removed to the amount of pollutants in the raw wastewater. In mathematical form, this is

$$\text{efficiency} = \frac{P_{\text{in}} - P_{\text{out}}}{P_{\text{in}}} \times 100 \tag{3-2}$$

where P_{in} = concentration of pollutant flowing into the treatment system

P_{out} = concentration of pollutant flowing out of the treatment system

Example 3.3

A sewage treatment plant influent has an average TSS concentration of 250 mg/L. If the average effluent TSS concentration is 20 mg/L, what is the removal efficiency for TSS? If the flow rate is 5 ML/d, how many kilograms of suspended solid is discharged in the plant effluent each day?

Solution

First applying Equation 3-2, we get

$$\text{efficiency} = \frac{250-20}{250} \times 100 = \frac{230}{250} \times 100 = 92\%$$

$$\text{Kilograms per day} = Q \times C = 5\ \text{ML/d} \times 20\ \text{mg/L} = 100\ \text{kg/d}$$

3.3.2 Water Quality Requirements

In the past, when water was plentiful and development sparse, common law determined water rights. The rights of owners living on the banks (nonsalable riparian rights) allowed "reasonable" use of the water, provided downstream users were not adversely affected. Later, as development occurred, some authorities allocated salable water rights to industries, mining companies, and others for as long as the continued to use the water and whether they were riparian owners or not.

The simple approaches of the past, based on common law, are not adequate for highly developed societies with limited water resources that need to be protected from pollution. Control of water quality may be under federal jurisdiction, as in the United States and most other countries, or under a regional authority (for example, a state, province, or county), as in Canada. Whatever the basis, it is essential that standards or objectives (guidelines or criteria for meeting desirable goals) be established. The quality of either the receiving water or the effluent may be the controlling requirement. If as seems reasonable in theory, the receiving water governs, then administration and enforcement of water quality standards or criteria are necessary. The quality of municipal and industrial discharges must be related to the stream or lake standards and enforced, difficult though that may be. On the other hand, if the requirements regulate effluent quality, monitoring and control are much simpler in practice. Offsetting this advantage is the problem that treatment requirements are not related to the water quality of the stream or lake. The result may be more costly treatment than is warranted where ample dilution is available, but insufficient treatment where it is not. A combination of stream and effluent requirements may be the best compromise. Assimilation studies could establish allowable loadings to a stream or lake, and this capacity could be allocated to users on a priority basis. Anyone discharging wastewater to these waters would have to meet these requirements or minimum applicable effluent standards, whichever was least detrimental to the receiving water.

Under the *Clean Water Act* as amended in 1977, the United States has required since July 1,

1988, that all discharges receive the "best conventional pollution control technology" regardless of the assimilation capacity of the stream. This requirement, which implies secondary treatment, may, like the original plan for zero discharge of pollutants, be modified where ample dilution is available, the location is remote, essential industries are involved, needed employment is provided, or other politically justified reasons make it expedient to do so.

Most developed countries have effluent and ambient water quality requirements that vary depending upon the characteristics of the wastewater and whether the receiving waters will be used for water supply, recreation, irrigation, or industrial purposes. Examples of effluent and ambient water quality requirements for Ontario, the United States, and Japan are given in Table 3.2. The effluent requirements apply generally to municipal wastewater where at least 85 percent removal of organic matter is provided. The criteria for stream quality pertain to receiving waters that are either suitable (with filtration) as a raw water supply or acceptable for swimming.

Table 3.2 Effluent and water quality requirements

Parameter	Wastewater effluent	Stream quality
	Canadian Objectives (Ontario)	
BOD_5	15 mg/L max.	4 mg/L max.
SS	15 mg/L max.	—
DO	2 mg/L min.	4 mg/L min.
Total Coliforms	—	5,000/100 mL max. (water supply)
		1,000/100 mL max. (swimming area)
Fecal Coliforms	200/100 mL max.	500/100 mL max. (water supply)
	United States Standards (Typical)	
BOD_5	30 mg/L max.	4 mg/L max.
SS	30 mg/L max.	—
DO	—	4 mg/L min.
Total Coliforms	—	5,000/100 mL max. (water supply)
		1,000/100 mL max. (swimming area)
Fecal Coliforms	200/100 mL max.	500/100 mL max. (water supply)
	Japanese Standards	
BOD_5	20 mg/L max.	2 mg/L max.
SS	70 mg/L max.	25 mg/L max.
DO	—	7.5 mg/L min.
Total Coliforms	—	5,000/100 mL max. (water supply)
		1,000/100 mL max. (swimming area)
Fecal Coliforms	30/100 mL max.	

(Source: J. Glynn Henry and Gary W. Heinke, *Environmental Science And Engineering*, Prentice Hall, 1989, P421)

3.4 SOURCES OF WATER POLLUTION

3.4.1 Point Sources

The sources of water pollution can be grouped into two classes, that is, point sources and non-point sources. Domestic sewage and industrial wastes are called point sources because they are generally collected by a network of pipes or channels and conveyed to a single point of discharge into the receiving water. Domestic sewage consists of wastes from homes, schools, office buildings, and stores. The term municipal sewage is used to mean domestic sewage into which industrial wastes are also discharged. In general, point source pollution can be reduced or eliminated through proper wastewater treatment prior to discharge to a natural water body.

Domestic Sewage

Quantities of domestic sewage are commonly determined from water use. Because water is consumed by humans, only 70 to 90 percent of the water supplied reaches the sewers.

The composition of domestic sewage changes regularly in a day and the quantity of domestic sewage varies with the season. Comparing with the industrial wastewater, the domestic sewage has the following properties:

Fresh, aerobic domestic sewage has the odor of kerosene or freshly turned earth. Aged, septic sewage is considerably more offensive to the olfactory nerve. The characteristic rotten-egg odor of hydrogen sulfide and the mercaptans is indicative of septic sewage. Fresh sewage is typically gray in color. Septic sewage is black.

Domestic sewage temperatures normally range between 10 and 20 °C. The suspended solids of domestic sewage are relatively low, generally 200-500 mg/L. Domestic sewage belongs to low organic concentration wastewater. Its BOD_5 is about 210-600 mg/L and the pH value is between 7.2-7.6. Domestic sewage contains more nutrients and a variety of microorganisms, including pathogens.

The quality of municipal wastewater varies with the proportion of residential, commercial, and industrial contributors and the nature of the industrial wastes which the system receives.

Industrial Wastewater

Wastewaters from industries include employees sanitary wastes, process wastes from manufacturing, wash waters, and relatively uncontaminated water from heating and cooling operations.

Industrial processes generate a wide variety of wastewater pollutants.The characteristics and levels of pollutants vary significantly from industry to industry. In general, industrial wastewater has the following properties:

Some industrial wastewater temperatures can exceed 40 °C and can bring heat pollution when they are discharged into water. The suspended solids of industrial wastewater are high, which can run up to 30,000 mg/L. The COD is approximately between 400 to 10,000 mg/L and the pH value changes dramatically, generally 2-13. Some industrial wastewater is combustible because of the volatile liquid with low ignition point, such as gasoline, benzene etc.. Meanwhile, industrial wastewater often contains a variety of harmful ingredients. For example, there are many pollutants, such as mercury, lead, arsenic, cyanide, sulfide, naphthalene, benzene, nitro-compounds, acid, alkali etc. in chemical plant wastewater.

The wastewaters from processing are the major concern. They vary widely with the type of industry. In some cases, pretreatment to remove certain contaminants or equalization to reduce hydraulic load may be mandatory before the wastewater can be accepted into the municipal system. In contrast with the relatively consistent characteristics of domestic sewage, industrial wastewaters, often have quite different characteristics, even for similar industries. For this reason, extensive studies may be necessary to assess pretreatment requirements and the effect of the wastewater on biological processes.

Wastes are specific for each industry and can range from strong (high BOD_5) biodegradable wastes like those from meat packing, through wastes such as those from plating shops and textile mills, which may be inorganic and toxic and require on-site physical-chemical treatment before discharge to the municipal system.

3.4.2 Non-point Sources

Urban and agricultural runoff are characterized by multiple discharge points. These are called ***non-point sources***. Often the flow of polluted water flows over the surface of the land or along natural drainage channels to the nearest water body. Even when urban or agricultural runoff waters are collected in pipes or channels, they are generally transported the shortest possible distance for discharge so that wastewater treatment at each outlet is not economically feasible. Much of the non-point source pollution occurs during rain storms or spring snowmelt resulting in large flow rates which make treatment even more difficult. Reduction of non-point source pollution generally requires changes in land use practices and improved education.

3.5 PRINCIPLES OF WASTEWATER TREATMENT

Where small quantities of sewage discharge into relatively large rivers or water bodies, incidents of contaminated water supplies or hazards to public health are rare. This is because of the dilution of the contaminants and the natural purification that takes place. With increasing population, these mitigating factors become less effective, and eventually, some form of wastewater treatment is warranted. The necessary treatment efficiency can be related to the assimilative capacity of the receiving water, that is, its ability to accept organic matter, nitrogen, and other pollutants without creating problems.

The need for wastewater treatment arises in all countries. In the less developed regions, the treatment of domestic wastes for the protection of public health is still the principal concern. For example, in vast areas of India, Africa, and South America, untreated wastes enter receiving waters that are used directly by large populations for washing, bathing, and drinking. In most developed countries, the need is shifting from purely public health considerations to the control of eutrophication, protection of aquatic life, and concern over toxic substances in the environment.

The suspended, colloidal, and dissolved contaminants (both organic and inorganic) in wastewater may be removed physically, converted biologically, or changed chemically. Different hazardous wastes need different treatment methods. According to the principles, treatment processes are often categorized as being physical, chemical and biological.

3.5.1 Physical Processes

Physical processes include gravity separation, phase change systems, such as air and steam stripping of volatiles from liquid wastes, and various filtering operations, including carbon adsorption.

Sedimentation The simplest physical treatment systems that separate solids from liquids take advantage of gravity settling and natural flotation. Special sedimentation tanks and clarification tanks are designed to encourage solids to settle so they can be collected as a sludge from the bottom of the tank. Some solids will float naturally to the surface and they can be removed with a skimming device. It is also possible to encourage flotation by introducing finely divided bubbles into the waste stream. The bubbles collect particles as they rise and the combination can be skimmed from the surface. Seperated sludges can then be further concentrated by evaporation, filtration, or centrifugation.

Adsorption Physical treatment can also be used to remove small concentrations of hazardous substances dissolved in water that would never settle out. One of the most commonly used techniques for removing organics involves the process of adsorption, which is the physical adhesion of chemicals onto the surface of a solid. The most commonly used adsorbent is a very porous matrix of granular activated carbon(GAC), which has an enormous surface area. A single handful of GAC has an internal surface of about one acre.

Aeration For chemicals that are relatively volatile, another physical process, aeration, can be used to drive the contaminants out of solution. These stripping system typically use air, though in some circumstances steam in used. In the most commonly used air stripper, contaminated water is sprayed downward through packing material in a tower, while air is blown upward carrying away the volatiles with it. Such a packed-tower can easily remove over 95 percent of the volatile organic compounds. Another type of stripper, called an induced-draft tower,does not use a blower or packing material. In the induced-draft tower, a carefully engineered series of nozzles spray contaminated water horizontally through the sides of a chamber. Air passing through the chamber draws off the volatiles. Induced-draft strippers are cheaper to build and operate, but their performance is much lower than a packed-tower.

The volatiles removed in an air stripper are, in some circumstances, released directly to the atmosphere. When discharged into the atmosphere is unacceptable, a GAC treatment system can be added to the exhaust air, as shown in Figure 3.3.

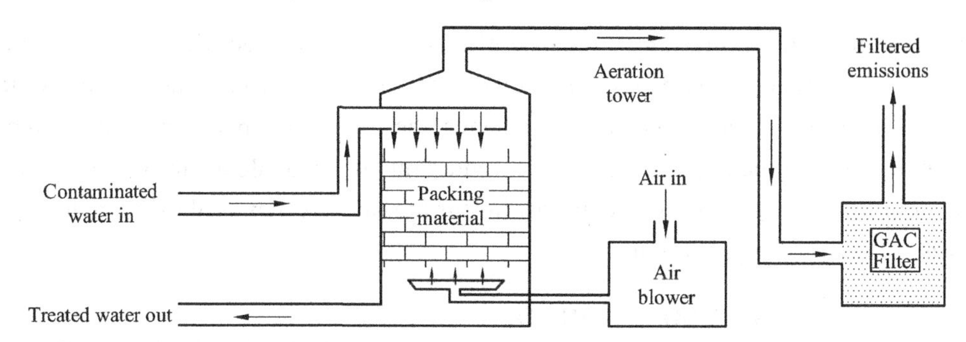

Figure 3.3 An air stripping tower followed by a granular activated carbon filter

(Source: J. Gilbert M. Masters, *Introduction to Environmental Engineering and Science*, Prentice Hall, 1991, P251)

3.5.2 Chemical Processes

Chemical treatment transforms waste into less hazardous substances using such techniques as pH neutralization, oxidation or reduction, and precipitation. Chemically treating hazardous waste not only has the potential advantage of converting it to less hazardous forms, but can also produce useful by-products in some circumstances.

Neutralization Wastewater, having a pH of less than 2 or more than 12.5, is corrosive. Such wastewater can be chemically neutralized. Acidic wastewaters are usually neutralized with slaked lime [$Ca(OH)_2$] in a continuously stirred chemical reactor. The rate of addition of lime is controlled with a feedback control system that monitors pH and adjusts the feed rate accordingly.

Alkaline wastwaters may be neutralized by adding acid directly or by bubbling in gaseous CO_2, forming carbon acid (H_2CO_3). The advantage of CO_2 is that it is quite often readily available in the exhaust gas from any combustion process at the treatment site.

Chemical Precipitation Chemical precipitation is a common method for removing heavy metals from wastewater by adjusting pH. By properly changing pH, the solubility of toxic metals can be decreased, leading to formation of a precipitate that can be removed by settling and filtration.

Frequently, the precipitation involves the use of lime,$Ca(OH)_2$, or caustic (NaOH) to form metal hydroxides. For example, the following reaction suggests the use of lime to form the hydroxide of a divalent metal (M^{2+}):

$$M^{2+} + Ca(OH)_2 \longrightarrow M(OH)_2 + Ca^{2+} \quad (3\text{-}3)$$

Metal hydroxides are relatively insoluble in basic solutions. Some of them are amphoteric, that is, they have some pH at which their solubility is a minimum. Since each metal has its own optimum pH, it is tricky to control precipitation of a mix of different metals in the same wastewater.

Chemical Reduction-Oxidation When electrons are removed from an ion, atom, or molecule, the substance is oxidized; when electrons are added, it is reduced. Both oxidation and reduction occur in the same reaction; hence the abbreviation redox. One of the most important redox treatment processes is the reduction of hexavalent chromium to trivalent chromium in large electroplating operations. Sulfur dioxide is often used as the reducing agent, as shown in the following reactions:

$$3SO_2 + 3H_2O \longrightarrow 3H_2SO_3 \quad (3\text{-}4)$$

$$2CrO_3 + 3H_2SO_3 \longrightarrow Cr_2(SO_4)_3 + 3H_2O \quad (3\text{-}5)$$

The trivalent chromium formed in reaction (3-5) is much less toxic and more easily precipitated than the original hexavalent chromium. Notice that the chromium in reaction (3-5) is reduced from an oxidation state of +6 to +3, while the sulfur is oxidized from +4 to +6.

Chemical treatment alone or with other processes is frequently necessary for industrial wastes that are not amenable to treatment by biological means. The oxidation of hexavalent chromium to the nontoxic trivalent form in the disposal of plating wastes is an example. Chemical processes are also useful in municipal waste treatment, where phosphorus concentrations are reduced and removal of solids increased by precipitation of these contaminants with metallic salts.

3.5.3 Biological Processes

Most of the organic constituents in wastewater can serve as food to provide energy for microbial growth. This is the principle used in biological waste treatment, where organic substrate is converted by microorganisms, mainly bacteria (with the help of protozoa), to carbon dioxide, water and more new cells. The microorganisms may be aerobic (requiring free oxygen), anaerobic (not requiring free oxygen), or facultative (growing with or without oxygen). Processes in which microorganisms use bound oxygen (from NO_3^- for denitrification, for example) are often called anoxic rather than anaerobic. The microbial population may be maintained in the liquid as suspended growth, referred to as mixed liquor suspended solids or volatile suspended solids (MLSS or MLVSS), or it may be attached to some medium in a fixed-film process. The rate of microbial growth varies directly with the amount of available substrate. In a batch culture when food is not limiting, the microbial population, after an initial lag period, grows rapidly at a logarithmic rate. As food decreases, growth slows until, at some point, growth stops and the number of new cells produced is balanced by the number of old cells which are dying. When the substrate is exhausted, the number of microorganisms declines as old cells decompose (lyse) releasing their nutrients for use by new microorganisms. These four phases referred to as the lag(A), log growth (B), declining growth (C) and endogenous (auto-oxidation) phase (D) are shown schematically in Figure 3.4.

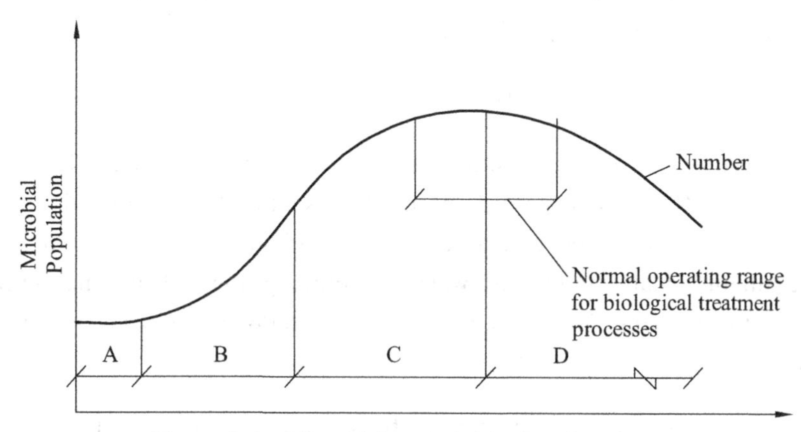

Figure 3.4 Microbial growth in a batch culture

(Source: J. Glynn Henry and Gary W. Heinke, *Environmental Science And Engineering*, Prentice Hall, 1989, P432)

In a continuous biological process, the system normally operates at some point on the growth curve toward the end of the declining growth phase, or into the endogenous phase where cells utilize their own protoplasm to obtain energy. Utilization of the logarithmic growth phase in wastewater treatment has been impractical, because substrate removal is incomplete and no economical way has been found to separate the microbial population from the liquid.

Aerobic/Anoxic Processes In aerobic processes (i.e. molecular oxygen is present) heterotrophic bacteria (those obtaining carbon from organic compounds) oxidize about one third of the colloidal and dissolved organic matter to stable end products (CO_2+H_2O) and convert the remaining two-thirds into new microbial cells that can be removed from the wastewater by settling. The overall biological conversion proceeds sequentially, with oxidation of carbonaceous material as the first step, i.e.,

$$\text{Organic matter} + O_2 \longrightarrow CO_2 + H_2O + \text{new cells} \tag{3-6}$$

Under continuing aerobic conditions, autotrophic bacteria (those obtaining carbon from inorganic compounds) then convert the nitrogen in organic compounds to nitrates according to the simplified equations

$$\text{Organic N} \longrightarrow NH_3 \text{ (decomposition)} \tag{3-7}$$

and

$$NH_3 + O_2 \xrightarrow[\text{bacteria}]{\text{nitrifying}} NO_2^- \longrightarrow NO_3^- \text{(nitrification)} \tag{3-8}$$

No further changes in the nitrates take place unless the process becomes anoxic (i.e. only bound oxygen is present). Under such conditions, heterotrophic bacteria convert the nitrates to odorless nitrogen gas:

$$NO_3^- \xrightarrow[\text{bacteria}]{\text{denitrifying}} NO_2^- \longrightarrow N_2 \text{ (denitrification)} \tag{3-9}$$

Under continuing anoxic conditions, any sulfates present are reduced to odorous hydrogen sulfide gas according to the descriptive equation

$$SO_4^{2-} \xrightarrow[\text{bacteria}]{\text{sulfate reducing}} H_2S \tag{3-10}$$

The preceding naturally occurring reactions are used in various biological processes for the treatment of wastewater. In every case, nutrients essential for biological growth must be present in the waste or must be added. Unlike municipal wastewater, which contains the necessary ingredients, many industrial wastes, including those from the pulp and paper, cannery, and meat processing industries, need nitrogen and/or phosphorus added for biological growth to occur.

Ever since the importance of wastewater treatment became recognized, municipalities and industries have relied almost exclusively on aerobic rather than anaerobic biological processes for treating their liquid organic wastes. Aerobic treatment has predominated because of its simplicity, stability, efficient and rapid conversion of organic contaminants to microbial cells, and relatively odor-free operation.

Although all aerobic biological oxidation processes use microorganisms to convert organic contaminants, the methods for accomplishing this conversion are varied and numerous.

Anaerobic Processes In anaerobic biological processes (i.e. no oxygen is present), two groups of heterotrophic bacteria, in a two-step liquefaction/gasification process, convert over 90 percent of the organic matter present, initially to intermediates (partially stabilized end products including organic acids and alcohols) and then to methane and carbon dioxide gas:

$$\text{Organic matter} \xrightarrow[\text{bacteria}]{\text{acid-forming}} \text{intermediates} + CO_2 + H_2S + H_2O \tag{3-11}$$

$$\text{Organic acids} \xrightarrow[\text{bacteria}]{\text{methane}} CH_4 + CO_2 \tag{3-12}$$

The process is universally used in heated anaerobic digesters, where primary and biological sludges are retained for approximately 30 days at 35 °C to reduce their volume and their putrescibility, and thus simplify their disposal, usually on agricultural land.

Two major advantages of anaerobic processes over aerobic ones are, that they provide useful energy in the form of methane and that sludge production is only about 10 percent of that from aerobic processes for converting the same amount of organic matter. This is advantageous in the treatment of high-strength wastes, where the handling of large volumes of sludge would be a problem.

3.6 MUNICIPAL WASTEWATER TREATMENT SYSTEM

Municipal sewage refers to the sum of all sewage discharged into the urban sewage networks. It mainly comes from domestic sewage and industrial wastewater. The objective of municipal wastewater treatment is to reduce the concentrations of specific pollutants to the level at which the discharge of the effluent will not adversely affect the receiving body. The basic components that accomplish all of these in a municipal wastewater treatment plant are shown in Figure 3.5.

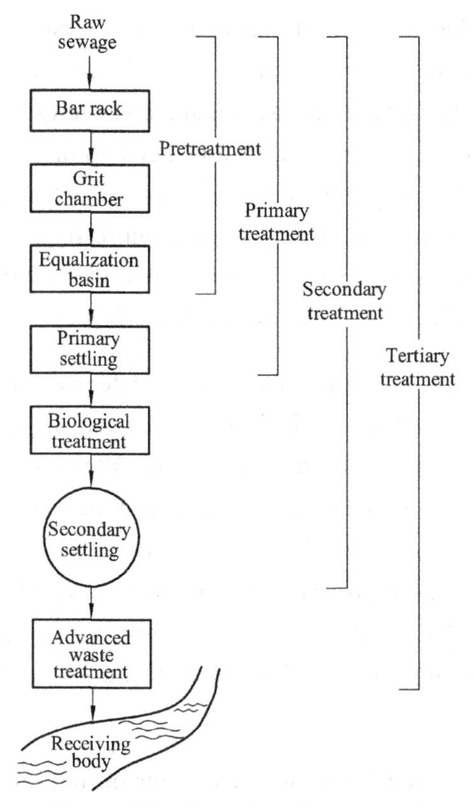

Figure 3.5 Degrees of treatment

(Source: Mackenzie L. Davis and David A. Cornwell, *Introduction to Environmental Engineering*, Tsinghua University Press, 2007, P439)

It is commonly assumed that each of the "degrees of treatment" noted in Figure 3-5 includes the previous steps. For example, primary treatment is assumed to include the pretreatment process: bar rack, grit chamber, and equalization basin. Likewise, secondary treatment is assumed to include all the processes of primary treatment: bar rack, grit chamber, equalization basin, and primary settling tank.

The purpose of pretreatment is to remove large objects, such as animal carcasses and chunk of debris, to protect the wastewater treatment plant (WWTP) equipment that follows. In some older municipal plants the equalization step may not be included.

The major goal of primary treatment is to remove from wastewater those pollutants which will either settle or float. Primary treatment systems are usually physical processes. Primary treatment will typically remove about 60 percent of the raw sewage suspended solids and 35 percent of the BOD_5. Soluble pollutants are not removed. At one time, this was the only treatment used by many cities. Now the law in many countries requires that municipalities provide secondary treatment. Although primary treatment alone is no longer acceptable, it is still frequently used as the first treatment step in a secondary treatment system.

The major goal of secondary treatment is to remove the soluble BOD_5 that escapes the primary process and to provide added removal of suspended solids. Secondary treatment is typically achieved by using biological processes. These provide the same biological reactions that would occur in the receiving water if it had adequate capacity to assimilate the wastewater. The secondary treatment processes are designed to speed up these natural processes so that the breakdown of the degradable organic pollutants can be achieved in relatively short time periods. Although secondary treatment may remove more than 85 percent of the BOD_5 and suspended solids, it does not remove significant amounts of nitrogen, phosphorus, or heavy metals, nor does it completely remove pathogenic bacteria and viruses.

In cases where secondary levels of treatment are not adequate, new treatment processes are applied to the secondary effluent to provide advanced wastewater treatment (AWT). Some of these processes may involve physical treatment (e.g., filtration to remove solids), chemical treatment (e.g., precipitation to remove phosphorus), or biological treatment (e.g., constructed wetland to remove BOD) – much like adding a typical water treatment plant to the tail end of a secondary plant. Some of these processes can remove as much as 99 percent of the BOD_5 and phosphorus, all suspended solids and bacteria, and 95 percent of the nitrogen. They can produce a sparkling clean, colorless, odorless effluent indistinguishable in appearance from a high-quality drinking water.

Most of the impurities removed from the wastewater do not simply vanish. Some organics are broken down into harmless carbon dioxide and water. Most of the impurities are removed from the wastewater as a solid, that is, sludge. Because most of the impurities removed from the wastewater are present in the sludge, sludge handling and disposal must be carried out carefully to achieve satisfactory pollution control.

3.6.1 Pretreatment of Municipal Wastewater

Several devices and structures are placed upstream of the primary treatment operation to provide protection to the wastewater treatment plant (WWTP) equipment. These devices and structures are classified as pretreatment because they have little affect in reducing BOD_5.

Bar Racks

Typically, the first device encountered by the wastewater entering the plant is a bar rack (Figure 3.6). The primary purpose of the rack is to remove trash and large objects which would damage or foul pumps, valves, and other mechanical equipment. Rags, logs, and other objects, which find their way into the sewer are removed from the wastewater on the racks.

Figure 3.6 Bar rack, used in wastewater treatment

(Source: Mackenzie L. Davis and David A. Cornwell, *Introduction to Environmental Engineering*, Tsinghua University Press, 2007, P439)

In modern WWTPs, the racks are cleaned mechanically. The cleaning rakes are automatically activated when the racks become sufficiently clogged to raise the water level in front of the bar. Mechanically cleaned racks have openings ranging from 15 to 75 mm. Maximum channel approach velocities range from 0.6 to 1.0 m/s. Minimum velocities of 0.3 to 0.5 m/s are necessary to prevent grit accumulation. Two channels are provided to allow one to be taken out of service for cleaning and repair.

Grit Chambers

Inert dense material such as sand, broken glass, silt, and pebbles is called grit. If these materials are not removed from the wastewater, they abrade pumps and other mechanical devices, causing undue wear. In addition, they have a tendency to settle in corners and bends, reducing flow capacity and, ultimately, clogging pipes and channels.

There are three basic types of grit removal devices: velocity controlled, aerated, and constant level short-term sedimentation basins. We will discuss only the first two, as they are the most common.

Velocity controlled This type of grit chamber, also known as a horizontal-flow grit chamber, can be analyzed by means of the classical laws of sedimentation for discrete, nonflocculating particles. Liquid velocity control is achieved by placing a specially designed weir at the end of the channel. A minimum of two channels must be employed so that one can be out of service without shutting down the treatment plant. Cleaning may be either by mechanical devices or by hand. Mechanical cleaning is favored for plants having average flows over 0.04 m^3/s. Theoretical detention times are set at about one minute for average flows.

Aerated grit chambers The spiral roll of the aerated grit chamber liquid "drives" the grit into a hopper which is located under the air diffuser assembly (Figure 3.7). The shearing action of the air bubbles is supposed to strip the inert grit of much of the organic material which adheres to its surface.

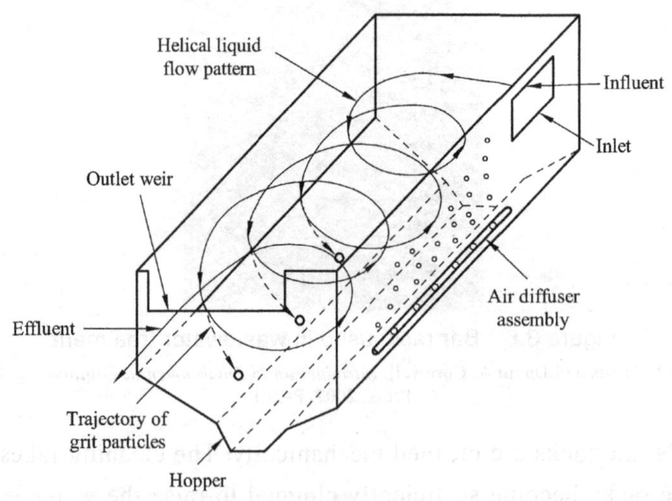

Figure 3.7 Aerated grit chamber

(Source: Meetcalf and Eddy, Inc., and G. Tchobanolous, *Wasterwater Engineering: Treatment Disposal Reuse*. New York: McGraw-Hill, 1979)

Aerated grit chamber performance is a function of the roll velocity and detention time. The roll velocity is controlled by adjusting the air feed rate. Nominal air flow values are in the range of 0.2 to 0. 5 cubic meters per minute of air per meter of tank length [$m^3/(min \cdot m)$]. Liquid detention times are usually set to be about 3 minutes at maximum flow. Length-to-width ratios range from 3:1 to 5:1 with depths on the order of 2 to 5 m.

Grit accumulation in the chamber varies greatly, depending on whether the sewer system is a combined type or a separate type, and on the efficiency of the chamber. For combined systems, 90 m^3 of grit per million cubic meters of sewage ($m^3/10^6\ m^3$) is not uncommon. In separate systems you might expect something less than 40 $m^3/10^6\ m^3$. Normally the grit is buried in sanitary landfill.

Comminutors

Devices that are used to macerate wastewater solid (rags, paper, plastic, and other materials) by revolving cutting bars are called comminutors. One common design is shown in Figure 3.8. Comminutors are most commonly used in small WWTPs with flow less than 0.2 m^3/s. These devices are placed downstream of the grit chambers to protect the cutting bars from abrasion. They are used as a replacement for the downstream bar rack but must be installed with a hand-cleaned rack in parallel in case they fail.

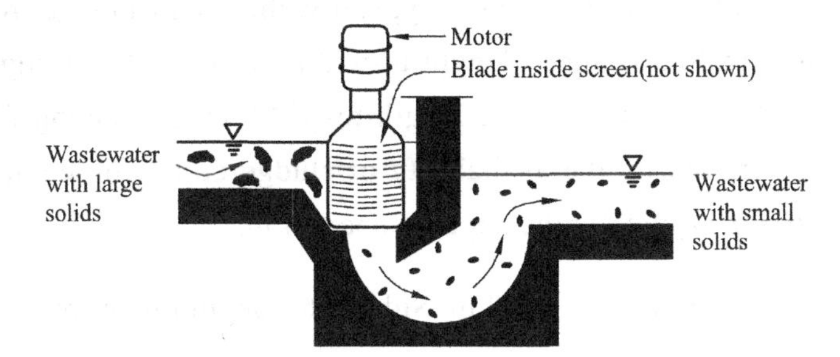

Figure 3.8 A typical comminutor

(Source: P. Aarne Vesilind and Susan M. Morgan, *Introduction to Environmental Engineering*, Thomson Brooks Cole, 2004, P247)

Equalization

Flow equalization is not a treatment process per se, but a technique that can be used to improve the effectiveness of both secondary and advanced wastewater treatment processes. Wastewater does not flow into a municipal wastewater treatment plant at a constant rate. The flow rate varies from hour to hour, reflecting the living habits of the area served. The constantly changing amount and strength of wastewater to be treated makes efficient process operation difficult. Also, many treatment units must be designed for the maximum flow conditions encountered, which actually results in their being oversized for average conditions. The purpose of flow equalization is to dampen these variations so that the wastewater can be treated at a nearly constant flow rate. Flow equalization can significantly improve the performance of an existing plant and increase its useful capacity. In new plants, flow equalization can reduce the size and cost of the treatment units.

Flow equalization is usually achieved by constructing large basins that collect and store the wastewater flow and from which the wastewater is pumped to the treatment plant at a constant rate. These basins are normally located near the head end of the treatment works, preferably downstream of pretreatment facilities such as bar screens, grit chambers, and comminutors. Adequate aeration and mixing must be provided to prevent odors and solids deposition.

3.6.2 Primary Treatment of Municipal Wastewater

With the screening completed and the grit removed, the wastewater still contains light organic suspended solids, some of which can be removed from the sewage by gravity in a sedimentation tank. These tanks can be round or rectangular. The mass of settled solids is called ***raw sludge***. The sludge is removed from the sedimentation tank by mechanical scrapers and pumps. Floating materials, such as grease and oil, rise to the surface of the sedimentation tank, where they are collected by a surface skimming system and removed from the tank for further processing.

Rectangular tanks with common-wall construction are frequently chosen because they are advantageous for sites with space constraints. Typically, these tanks range from 15 to 100 m in length and 3 to 24 m in width. Common length-to-width ratios for the design of new facilities range from 3:1 to 5:1. Existing plants have length-to-width ratios ranging from 1.5:1 to 15:1. The width is often controlled by the availability of sludge collection equipment. Side water depth range from 3 to 5 m. Typically the depth is about 4 m.

Circular tanks have diameters from 3 to 60 m. Side water depth range from 3 to 5 m.

As in water treatment clarifier design, overflow rate is the controlling parameter for the design of primary settling tanks. At average flow, overflow rates typically range from 25 to 60 m/d. When waste-activited sludge is returned to the primary tank, a lower range of overflow rates is chosen (25 to 35 m/d). Under peak flow conditions, overflow rates may be in the range of 80 to 120 m/d.

Hydraulic detention time in the sedimentation basin ranges from 1.5 to 2.5 hours under average flow conditions. A 2.0-hour detention time is typical.
As mentioned previously, approximately 50 to 60 percent of the raw sewage suspended solids and as much as 30 to 35 percent of the raw sewage BOD_5 may be removed in the primary tank.

3.6.3 Secondary Treatment of Municipal Wastewater

The water leaving the primary clarifier has lost much of the suspended organic matter but still contains a high demand for oxygen due to the dissolved biodegradable organics. This demand for oxygen must be reduced (energy expended) if the discharge is not to create unacceptable conditions in the watercourse. The major purpose of secondary treatment is to remove the soluble BOD that escapes primary treatment and to provide further removal of suspended solids. Except in rare circumstances, almost all secondary treatment methods use microbial action to reduce the energy level (BOD) of the waste.

The basic differences among all these alternatives are how the waste is brought into contact with the microorganisms. Biological treatment in secondary phase takes place in either a suspended-growth system or a fixed-film process. In the former, the microorganisms are kept in suspension in the wastewater by mixing or aerating devices. Such devices must satisfy the requirements for both mixing and oxygen transfer. With attached growth systems, the active microorganisms grow on the surface of rock, plastic, or other medium with which the waste is brought into contact.

Suspended-Growth Systems

Conventional activated sludge The most widely used application of suspended growth is the conventional activated sludge process which is a biological wastewater treatment technique in which a mixture of wastewater and biological sludge (microorganisms) is agitated and aerated. The biological solids are subsequently separated from the treated wastewater and returned to the aeration process as needed.

The activated sludge process derives its name from the biological mass formed when air is continuously injected into the wastewater. Under such conditions, microorganisms are mixed thoroughly with the organics under conditions that stimulate their growth through use of the organics as food. As the microorganisms grow and are mixed by the agitation of the air, the individual organisms clump together (flocculate) to form an active mass of microbes (biologic floc) called ***activated sludge***.

The key to the **conventional** activated sludge system is the reuse of microorganisms. The system, shown as a block diagram in Figure 3.9, consists of an aeration tank full of waste liquid (from the primary clarifier) and a mass of microorganisms. Air is bubbled into this tank to provide the necessary oxygen for the survival of the aerobic organisms. The microorganisms come in contact with the dissolved organics and rapidly adsorb these organics on their surface. In time, the microorganisms decomposed this material to CO_2, H_2O, some stable compounds, and more microorganisms. The production of new organisms is relatively slow, and most of the aeration tank volume is in fact used for this purpose.

Once most of the food has been utilized, the microorganisms are separated from the liquid in a settling tank which is referred to as ***a secondary clarifier*** or ***final clarifier*** to differentiate it from the sedimentation basin used for primary settling. The liquid escapes over a weir and may be discharged into the recipient. The separation of microorganisms is an important part of the system. In the settling tanks, the microorganisms exist without additional food and become hungry.

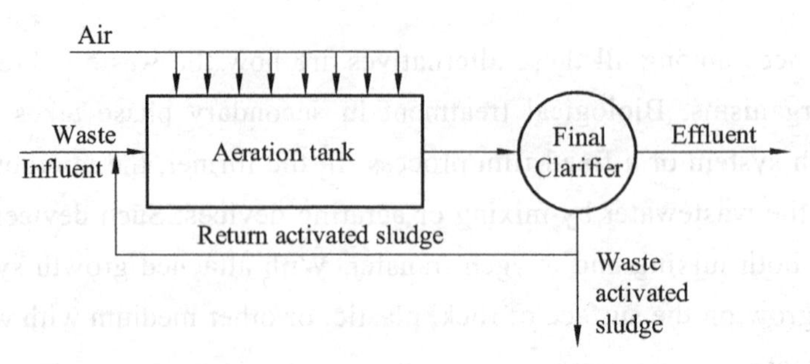

Figure 3.9 Block diagram of the activated sludge system

(Source: P. Aarne Vesilind J. Jeffrey Peirce and Ruth F. Weiner, *Environmental Engineering*, Butterworths, 1988, P157)

The settled microorganisms, now known as return activated sludge, are pumped to the head of the aeration tank where they find more food (organics in the effluent from the primary clarifier) and the process starts all over again. The volume of sludge returned to the aeration basin is typically 20 to 30 percent of the wastewater flow. The **conventional** activated sludge process is a continuous operation, with continuous sludge pumping the clean water discharge.

The activated sludge process is controlled by wasting a portion of the microorganisms each day in order to maintain the proper amount of microorganisms to efficiently degrade the BOD_5. Wasting means that a portion of the microorganisms is discarded from the process. The discarded microorgamisms are called ***waste activated sludge***. A balance is then achieved between growth of new organisms and their removal by wasting. If too much sludge is wasted, the concentration of microorganisms in the mixed liquor will become too low for effective treatment. If too little sludge is wasted, a large concentration of microorganisms will accumulate and, ultimately, overflow the secondary tank and flow into the receiving stream.

Conventional activated sludge systems are designed on the basis of loading , or the amount of organic matter (food) added relative to the microorganisms available. This ratio is known as the food-to-microorganisms ratio (F/M) and is a major design parameter. Unfortunately it is difficult to measure either F or M accurately, and engineers have approximated these by BOD and the SS in the aeration tank, respectively. The combination of the liquid and microorganisms undergoing aeration is known as mixed liquor, and thus the SS are called mixed liquor suspended solids (MLSS). The ratio of incoming BOD to MLSS, the

F/M ratio, is also known as the loading on the system, calculated as kilograms of BOD/day per kilograms of MLSS in the aeration tank.

Many variations of the conventional activated sludge process have been developed to address specific treatment problems. Differences in aeration time, MLSS concentration, solids retention time, and loading are some of the distinguishing characteristics of the processes, making them more suitable for one application than another.

Extended aeration The flow diagram for the extended aeration process is similar to that for conventional activated sludge, but with no primary tanks. In extended aeration system, the F/M ratio is low (little food for many microorganisms) and the aeration period (detention time in the aeration tank) is long. The microorganisms make maximum use of available food, resulting in a high degree of treatment. Such systems are widely used for isolated sources. An added advantage of extended aeration is that the ecology with the aeration tank is quite diverse and little excess biomass is created, resulting in little or no waste activated sludge to be disposed of – a significant saving in operating costs.

Tapered aeration and step aeration When the microorganisms first come in contact with the food, the process requires a great deal of oxygen. Accordingly, the DO level in the aeration tank drops immediately after the point at which the waste is introduced. If DO levels are measured over the length of a tank, extremely low concentrations are often found at the influent end of the aeration tank. These low levels of DO may be detrimental to the microbial population. Accordingly, two variations of the conventional activated sludge treatment have found some use: tapered aeration and step aeration (Figure 3.10). The former method consists of blasting additional air where needed, whereas step aeration involves the introduction of the waste at several locations, thus evening out the initial oxygen demand.

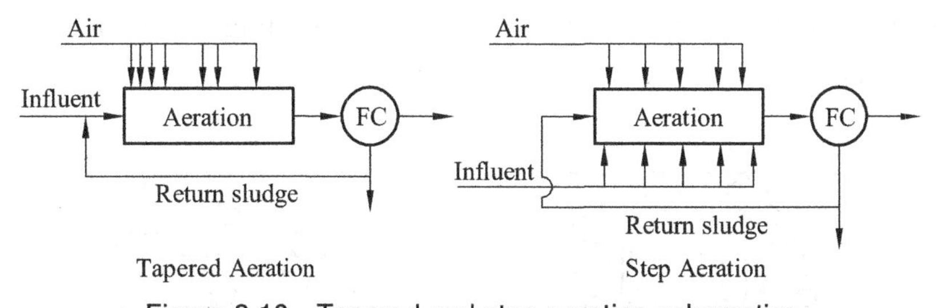

Figure 3.10 Tapered and step aeration schematics

(Source: P. Aarne Vesilind J. Jeffrey Peirce and Ruth F. Weiner, *Environmental Engineering*, Butterworths, 1988, P158)

Contact stabilization The contact stabilization, or biosorption, is a process in which the sorption and bacterial growth phases are separated by a settling tank. The advantage is that the growth can be achieved at high solids concentrations, thus saving tank space. Many existing activated sludge plants can be converted to biosorption plants when tank volume limits treatment efficiency. Figure 3.11 is a diagram of the biosorption process.

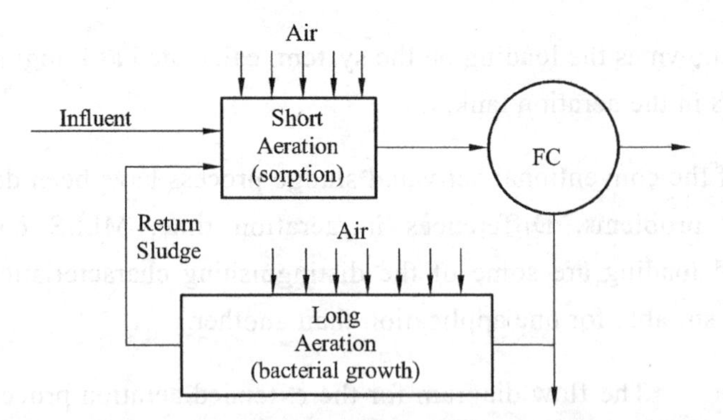

Figure 3.11 The biosorption modification of the activated sludge process

(Source: P. Aarne Vesilind J. Jeffrey Peirce and Ruth F. Weiner, *Environmental Engineering*, Butterworths, 1988, P159)

Oxygen-activated sludge The oxygen-activated sludge process, in which pure oxygen, rather than air, is used in covered tanks, was developed by Union Carbide in the late 1960s. It has some advantages over air-activated sludge. It can operate at higher loadings, thereby reducing aeration tank size, and can accept wider fluctuations in load and stronger wastes than can air-activated sludge. Process control of oxygen requirements by pressure sensors is simple and reliable. Where a high rate of oxygen transfer is necessary, as with high-strength wastes, the oxygen system has an advantage over the air-activated sludge with its limited transfer capability. Another situation where the oxygen system is more likely to have an economic advantage is where odor control is important and covering of the aeration tanks is intended. Also, control of odors will be much simpler since the volume of gas to be vented is only about 1 percent of that for air-activated sludge. However, the high cost of providing oxygen and the skills necessary to operate the process make it uncompetitive with the conventional activated sludge except for large plants.

The two principal means of introducing sufficient oxygen into the aeration tank are by bubbling compressed air through porous diffusers or beating air in mechanically. Both diffused air and mechanical aeration are shown in Figure 3.12.

Figure 3.12 (a)diffused aeration (b) mechanical aeration

(Source: P. Aarne Vesilind J. Jeffrey Peirce and Ruth F. Weiner, *Environmental Engineering*, Butterworths, 1988, P160)

The success of the activated sludge system depends on many factors. Of critical importance is the separation of the microorganisms in the final clarifier. The microorganisms in the system are sometimes very difficult to settle out, and the sludge is said to be a bulking sludge. Often this condition is characterized by a biomass composed almost totally of filamentous organisms that form a kind of lattice structure with the filaments and refuse to settle.

Treatment plant operators should keep a close watch on settling characteristics because a trend toward poor settling may be the forerunner of a badly upset (and hence ineffective) plant. The settleability of activated sludge is most often described by the sludge volume index (SVI), which is determined by measuring the milliliters of volume occupied by a sludge after settling for 30 minutes in a 1-L cylinder, and calculated as

$$\mathrm{SVI}=\frac{(\text{volume of sludge after 30min,in mL})\times 1{,}000}{\text{mg/L} \quad \text{of suspended solids}}$$

Example 3.4

A sample of mixed liquor was found to have SS = 4,000 mg/L and, after settling for 30 minutes in a 1-L cylinder, occupied 400 mL. Calculate the SVI.

$$\mathrm{SVI}=\frac{400\times 1{,}000}{4{,}000}=100$$

SVI values below 100 are usually considered acceptable, with SVIs greater than 200 defined as badly bulking sludges.

The causes of poor settling (high SVI) are not always known, and hence the solutions are elusive. Wrong or variable F/M ratios, fluctuations in temperature, high concentrations of heavy metals, and deficiencies in nutrients in the incoming wastewater have all been blamed for bulking. Cures include chlorination, changes in air supply, and dosing with hydrogen peroxide (H_2O_2)to kill the filamentous microorganisms.

When the sludge does not settle, the return activated sludge becomes thin (low SS concentration) and thus the concentration of microorganisms in the aeration tank drops. This results in a higher F/M ratio (same food input, but fewer microorganisms) and a reduced BOD removal efficiency.

Fixed Film Processes

In the fixed, or attached growth, method of secondary treatment, wastewater is brought into contact with microorganisms attached to a solid medium such as rock, plastic, or sand. Trickling filters and rotating biological contactors, two of the more common processes, fall under this category.

Trickling filters Trickling filters were first used in 1893. A trickling filter consists of a bed of coarse material, such as stones, slats, or plastic materials (media), which act as "contact beds" where settled sewage is spread. A very active biological growth forms on the stones, and the organisms obtain their food from the wastewater dripping through the bed of stones. Air is either forced through the stones or, more commonly, air circulation is obtained automatically by a temperature difference between the air in the bed and ambient temperature. In the older filters, the waste is sprayed onto the rocks from fixed nozzles. The newer designs utilize a rotating arm that moves under its own power, like a lawn sprinkler, distributing the waste evenly over the surface of the rocks(Figure 3.13). Often the flow is recirculated, thus obtaining a high degree of treatment. The most widely used design for many years was simply a bed of stones from 1 to 3m deep, and the rock filter diameters may range up to 60 m.

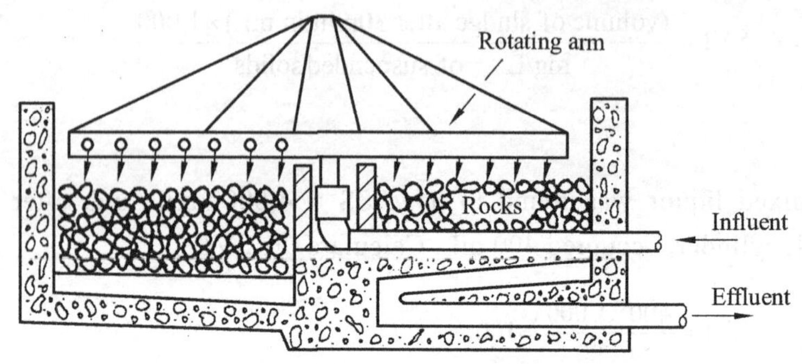

Figure 3.13 Trickling filter

(Source: P. Aarne Vesilind J. Jeffrey Peirce and Ruth F. Weiner, *Environmental Engineering*, Butterworths, 1988, P156)

Trickling filters are not primarily a filtering or straining process as the name implies. The rocks in a rock filter are 25 to 100 mm in diameter, and hence have openings too large to strain out solids. They are a means of providing large amounts of surface area where the microorganisms cling and grow in a slime on the rocks as they feed on the organic matter.

Excess growths of microorganisms wash from the rock media and would cause undesirably high levels of suspended solids in the plant effluent if not removed. Thus, the flow from the filter is passed through a sedimentation clarifier to allow these solids to settle out.

Although rock trickling filters have performed well for years, they have certain limitations. Under high organic loadings, the slime growths can be so prolific that they plug the void spaces between the rocks, causing flooding and failure of the system. Also, the volume of void spaces is limited in a rock filter, which restricts the circulation of air in the filter and the amount of oxygen available for the microbes. This limitation, in turn, restricts the amount of wastewater that can be processed.

To overcome these limitations, other materials have become popular for filling the trickling filter. These materials include modules of corrugated plastic sheets and plastic rings. These media offer larger surface areas for slime growths (typically 90 square meters of surface area per cubic meter of bulk volume, as compared to 40 to 60 square meters per cubic meter for 75 mm rocks) and greatly increase void ratios for increased air flow. The materials are also much lighter than rock (by a factor of about 30), so that the trickling filters can be much taller without facing structural problems. While rock in filters is usually not more than 3 m deep, synthetic media depths may reach 12 m thus reducing the overall space requirements for the trickling-filter portion of the treatment plant.

Trickling filters are classified according to the applied hydraulic and organic load. The hydraulic load may be expressed as cubic meters of wastewater applied per day per square meter of bulk filter surface area [$m^3/(d \cdot m^2)$] or, preferably, as the depth of water applied per unit of time (mm/s or m/d). Organic loading is expressed as kilograms of BOD_5 per day per cubic meter of bulk filter volume [$kg/(d \cdot m^3)$].

An important element in trickling filter design is the provision for return of a portion of the effluent to flow through the filter. This practice is called recirculation. The ratio of the returned flow to the incoming flow is called the recirculation ratio(r). Recirculation is practiced in stone filters for the following reasons:

1. To increase contact efficiency by bringing the waste into contact more than once with active biological material.

2. To dampen variations in loadings over a 24-hour period. The strength of the recirculated flow lags behind that of the incoming wastewater. Thus recirculation dilutes strong influent and supplements weak influent.

3. To raise the DO of the influent.

4. To improve distribution over the surface, thus reducing the tendency to clog and also reduce filter flies.

5. To prevent the biological slimes from drying out and dying during night time periods when flows may be too low to keep the filter wet continuously.

Recirculation may or may not improve treatment efficiency. The more dilute the incoming wastewater is, the less likely it is that recirculation will improve efficiency.

Recirculation is practiced for plastic media to provide the desired wetting rate to keep the microorganisms alive. Generally, increasing the hydraulic loading above the minimum wetting rate does not increase BOD_5 removal. The minimum wetting rate normally falls in the range of 25 to 60 m/d.

Rotating Biological Contactors(RBCs) Rotating biological contactors were first installed in the United States in 1969. An RBC process consists of a series of closely spaced, circular, plastic discs, that are typically 3 to 3.5 m in diameter and attached to a horizontal shaft and rotated, while about one-half of their surface area is immersed in wastewater (Figure

3.14). The speed of rotation of the discs is adjustable.

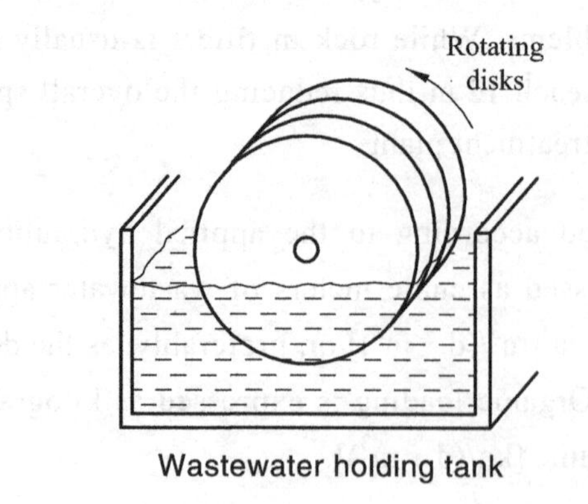

(a) RBC cross section

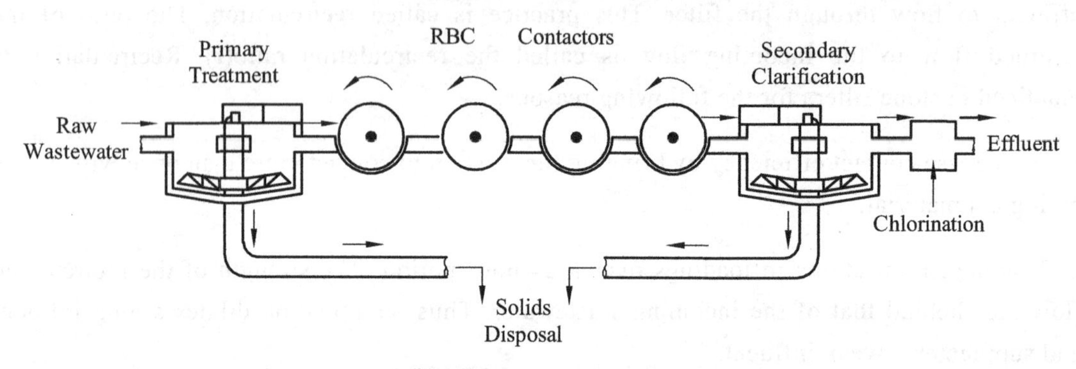

(b) RBC treatment system

Figure 3.14 RBC cross section and treatment system

(Source: Mackenzie L. Davis and David A. Cornwell, *Introduction to Environmental Engineering*, PWS Engineering, 1985, P295)

When the process is placed in operation, the microbes in the wastewater begin to adhere to the rotating surfaces and grow there until the entire surface area of the discs is covered with a 1 to 3 mm layer of biological slime. As the discs rotate, they carry a film of wastewater into the air, where it trickles down the surface of the discs, absorbing oxygen. As the discs complete their rotation, this film mixes with the reservoir of wastewater, adding to the oxygen in the reservoir and mixing the treated and partially treated wastewater. As the attached microbes pass through the reservoir, they absorb other organics for breakdown. The excess growth of microbes is sheared from the discs as they move through the reservoir. These dislodged organisms are kept in suspension by the moving discs. Thus, the discs serve several purposes:

1. They provide media for the buildup of attached microbial growth.

2. They bring the growth into contact with the wastewater.

3. They aerate the wastewater and the suspended microbial growth in the reservoir.

The attached growths are similar in concept to a trickling filter, except the microbes are passed through the wastewater rather than the wastewater passing over the microbes. Some of the advantages of both the trickling filter and activated sludge processes are realized.

As the treated wastewater flows from the reservoir below the discs, it carries the suspended growths out to a downstream settling basin for removal. The process can achieve secondary effluent quality or better. By placing several sets of discs in series, it is possible to achieve even higher degrees of treatment, including biological conversion of ammonia to nitrates.

3.6.4 Disinfection

The last treatment step in a secondary plant is the addition of a disinfectant to the treated wastewater in order to reduce further the possibility of disease transmission. The addition of chlorine gas or some other form of chlorine is the process most commonly used for wastewater disinfection. Often chlorine is used for the disinfection because it is fairly inexpensive. Chlorination occurs in simple holding basins designed to act as plug-flow reactors (Figure 3.15). The chlorine is injected at the beginning of the tank, and it is assumed that all the flow is in contact with the chlorine for 30 minutes. Prior to discharge, excess chlorine, which is toxic to many aquatic organisms, must be removed through dechlorination. The most common dechlorination method is bubbling in sulfur dioxide; the chlorine is reduced while the SO_2 is oxidized to sulfate. At this point the flow can be discharged into a receiving stream or other watercourse.

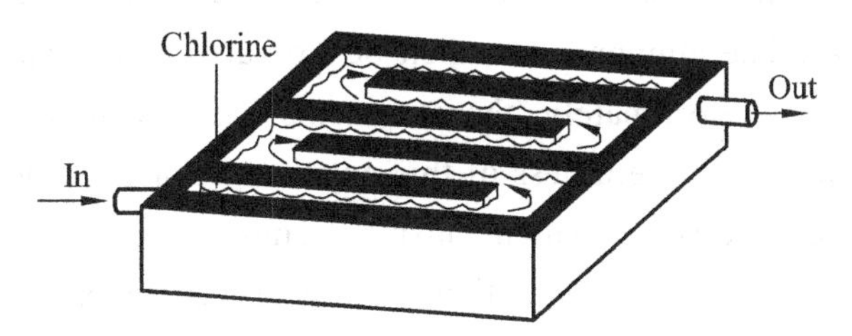

Figure 3.15 Plug-flow reactor for chlorination

(Source: P. Aarne Vesilind and Susan M. Morgan, *Introduction to Environmental Engineering*, Thomson Brooks Cole, 2004, P274)

Chlorination does not seem to make much sense from the ecological standpoint because the effluent must be assimilated into the aquatic ecology, and dosing wastewater treatment plant effluents with chlorine results in the production of chlorinated organic compounds, such as chloroform, a carcinogen. In addition, there is no epidemiological evidence that unchlorinated treatment plant effluents cause any public health problems. A compromise is to eliminate chlorination and to introduce other methods of disinfection, such as ultraviolet radiation and ozone. In some sensitive areas these techniques are already being used and with time may eventually eliminate chlorination of wastewater plant effluents.

3.6.5 Advanced Treatment of Municipal Wastewater

Although secondary treatment processes, when coupled with disinfection, may remove over 85 percent of the BOD and suspended solids and nearly all pathogens, only minor removal of some pollutants, such as nitrogen, phosphorus, soluble COD, and heavy metals, is achieved. In some circumstances, these pollutants may be of major concern. In these cases, processes capable of removing pollutants not adequately removed by secondary treatment are used in what is called ***tertiary wastewater treatment***, or ***advanced wastewater treatment*** (AWT). Many of the early advanced treatment facilities were designed with the primary purpose of removing nitrogen and phosphorus (the principal nutrients responsible for eutrophication), as well as to more completely reduce BOD. The following sections describe available AWT processes.

Nitrogen Removal

Nitrogen exists in a variety of forms in wastewater. As bacteria decompose waste, nitrogen that was bound up in complex organic molecules is released as ammonia nitrogen. Ammonia, in turn, exists in water in two forms: as ammonium ion (NH_4^+), which is highly soluble, and as ammonia gas (NH_3), which is not. As pH increases, the equilibrium relationship between these two forms is driven toward the less soluble ammonia gas:

$$NH_4^+ + OH^- \longrightarrow NH_3 \uparrow + H_2O \tag{3-13}$$

One method of nitrogen removal, ammonia stripping, is based on this reaction. In this process, the pH of treated wastewater is raised to at least 10, typically with quick lime (CaO), to form dissolved ammonia gas. The ammonia can then be liberated in a stripping tower of the sort illustrated in Figure3-2. Unfortunately, these systems have been plagued by a number of problems that have limited the usefulness of this approach. For one, the lime reacts with CO_2 to form a calcium carbonate scale, which must be removed periodically from the stripping surfaces. This scaling can be so severe that the tower may eventually cease to function. Ammonia stripping is also less effective in cold water, in part because ammonia is more soluble in cold water, making it harder to strip, but also because towers can ice up. The process also has been criticized because it simply transfers the pollution problem from one medium to another, in this case from water to air, creating an additional burden on the atmosphere.

A second approach to nitrogen control utilizes aerobic bacteria to convert ammonia (NH_4^+) to (NO_3^-), which is nitrification, followed by an anaerobic stage where different bacteria convert nitrates to nitrogen gas (N_2), which is denitrification. The overall process then is referred to as ***nitrification/denitrification***.

The nitrification step actually occurs in two stages. Ammonia is converted to nitrites (NO_2^-) by Nitrosomonas, while Nitrobacter oxidize nitrites to nitrates. The combination of steps can be summarized by

$$NH_4^+ + 2O_2 \xrightarrow{\text{bacteria}} NO_3^- + 2H^+ + H_2O \tag{3-14}$$

Nitrification does not begin to be important until domestic wastewater is at least 5-8 days old. Thus, if this method of nitrogen control is to be used, the wastewater must be kept in the treatment plant for a much longer time than would normally be the case. Detention times of 15 days or more are typically required. If reaction (3-12) takes place in the treatment plant rather than in the receiving body of water, at least the oxygen demand for nitrification is satisfied. The nitrogen, however, remains in the effluent, and if the process were to stop here, that nitrogen could go on to contribute to unwanted algal growth. To avoid this, the denitrification step is required.

The second phase of the nitrification/denitrification process is anaerobic denitrification:

$$2NO_3^- + \text{organic matter} \xrightarrow{\text{bacteria}} N_2\uparrow + CO_2 + H_2O \tag{3-15}$$

which release harmless, elemental nitrogen gas. The energy to drive this reaction comes from the organic matter indicated in (3-13). Since this denitrification process occurs after waste treatment, there may not be enough organic material left in the waste stream to supply the necessary energy and an additional source, usually methanol (CH_3OH), must be provided.

Phosphorus Removal

Only about 30 percent of the phosphorus in municipal wastewater is removed during conventional primary and biological treatment. Since phosphorus is very often the limiting nutrient, its removal from the waste stream is especially important when eutrophication is a problem.

Phosphorus in wastewater exists in many forms, but all of it ends up as orthophosphate ($H_2PO_4^-$, HPO_4^{2-}, and PO_4^{3-}). Removing phosphates is most often accomplished by adding a coagulant, usually alum [$Al_2(SO_4)_3$] or lime [$Ca(OH)_2$]. The pertinent reaction involving alum is

$$Al_2(SO_4)_3 + 2PO_4^{3-} \longrightarrow 2AlPO_4\downarrow + 3SO_4^{2-} \tag{3-16}$$

Alum is sometimes added to the aeration tank when the activated sludge process is being used, thus minimizing the need for additional equipment.

The reaction for precipitation with lime can be represented as

$$5Ca(OH)_2 + 3HPO_4^{2-} \longrightarrow Ca_5OH(PO_4)_3\downarrow + 3H_2O + 6OH^- \tag{3-17}$$

Where the precipitate formed is called calcium hydroxyphosphate, or, hydroxylapatite. When lime is used as the coagulant, it is often used after biological treatment, especially when ammonia stripping is also part of the treatment process. The lime not only causes the phosphate to precipitate out of solution, it also raise the pH of the waste stream so that soluble ammonium ions are converted to ammonia gas.

Further Organic Removal

Oxidation ponds are commonly used for BOD removal. The oxidation, or polishing, pond is essentially a hole in the ground, a large pond used to confined the plant effluent before it is discharged into the natural watercourse. Such ponds are designed to be aerobic, and because light penetration for algal growth is important, a large surface area is needed.

Activated carbon adsorption is another method of BOD removal, but this process has the added advantage that inorganics as well as organics are removed. The mechanism of adsorption on activated carbon is both chemical and physical, with tiny crevices catching and holding colloidal and smaller particles. An activated carbon column is similar to an ion exchange column. It is a completely enclosed tube with dirty water pumped up from the bottom and the clear water exiting at the top. As the carbon becomes saturated with various materials, the dirty carbon must be removed from the column to be regenerated, or cleaned. Removal is often continuous, with clean carbon being added at the top of the column. The regeneration is usually done by heating the carbon in the absence of oxygen, driving off the organic matter. A slight loss in efficiency is noted with regeneration, and some virgin carbon must always be added to ensure effective performance.

3.6.6 Sludge treatment

In the process of purifying the wastewater, another problem is created: sludge. The sludge is made of materials settled from the raw wastewater and solids generated in the wastewater treatment processes.The higher the degree of wastewater treatment is, the larger the residue of sludge that must be handled. Satisfactory treatment and disposal of the sludge can be the single most complex and costly operation in a municipal wastewater treatment system. Sludge disposal usually accounts for 50% of the cost of running a plant.

Historically the first and still the most widely used method of making the sludge less objectionable is anaerobic digestion. The process, as the name implies, does not rely on the availability of free oxygen and progresses in two steps:

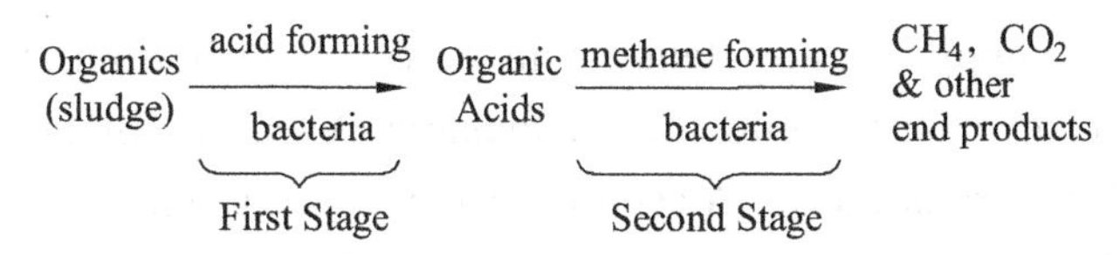

The methane formers are strict anaerobes unable to function in the presence of oxygen and very sensitive to environmental conditions such as temperature, pH, and toxins. If a digester goes "sour," the methane formers have been inhibited in some way, with a concomitant decrease in gas production, but the acid formers keep chugging a way, making more organic acids. This has the effect of further lowering the pH and making conditions even worse for the methane formers. A sick digester is therefore difficult to cure without massive doses of lime of other antacids.

Most treatment plants have two kinds of digesters – primary and secondary (Figure 3.16). The primary digester is covered, heated and mixed to increase the reaction rate. The temperature of the sludge is usually about 95 °F(35 °C). Secondary digesters are not mixed or heated and are used for storage of gas and for concentrating the sludge by settling. The liquid "supernatant" is pumped back to the main plant for further treatment. The cover of the secondary digester often floats up and down, depending on the amount of gas stored. The gas is high enough in methane to be used as a fuel, and is in fact used to heat the primary digester.

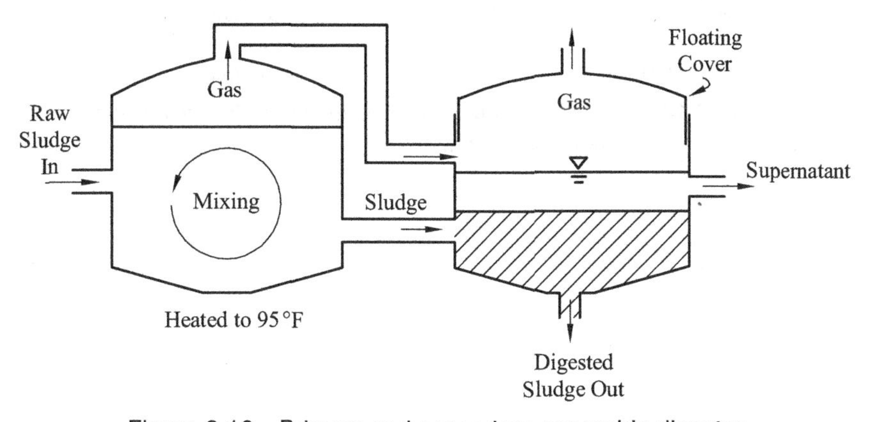

Figure 3.16 Primary and secondary anaerobic digester

(Source: P. Aarne Vesilind, *Environmental Pollution And Control*, Ann Arbor Science, 1975, P77)

Sludge which has been well digested (e.g., 20 days) and has a dark color does not smell badly. It can be pumped out of a digester for direct disposal on farmland. Although digested sludge has considerable value as a fertilizer and soil conditioner, most farmers do not want to bother with sludge, and thus other methods of disposal are necessary.

The sludges withdrawn from the treatment processes are still largely water, as much as 97 percent. Sludge treatment processes, then, are concerned with separating the large amounts of water from the solid residues to reduce the volume. The most popular methods of dewatering are sand drying beds, vacuum filters and centrifuges.

Sand drying beds are simple shallow tanks with sand on the bottom. The water drains out through the sand and evaporates into the atmosphere. Vacuum filters are large perforated drums wrapped with a porous fabric. A section of the drum is immersed in the sludge and a vacuum is pulled within the drum. This results in water being drawn through the cloth and a cake of solids being deposited on the cloth. The cake is then scraped off the rotating drum.

Centrifuges used in wastewater treatment are solid bowls which spin on a horizontal axis. Sludge is pumped inside the rotating bowl and the centrifugal force throws the solid particles to the inside wall. The solids are scraped out of the bowl by a screw conveyor while the liquid escapes over weirs on the opposite side of the bowl.

The vacuum filter and the centrifuge, shown schematically in Figure3.17, can dewater sludge to a condition dry enough to truck away. Although digested and dewatered sludge is a great soil conditioner and a reasonably good fertilizer, the usual method of disposal is into a convenient hole-in-the-ground.

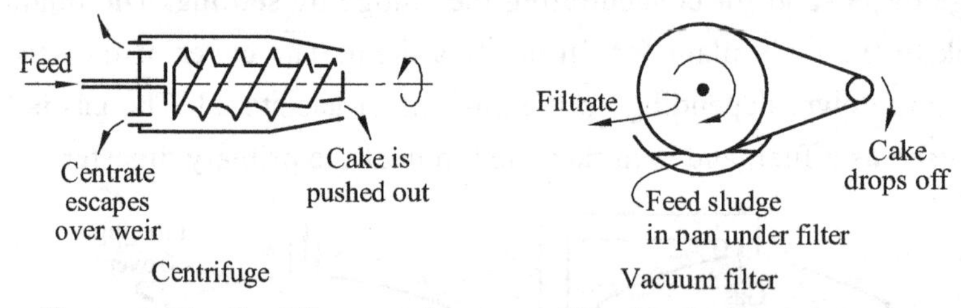

Figure 3.17 Centrifuge and vacuum filter for sludge dewatering

(Source: P. Aarne Vesilind, *Environmental Pollution And Control*, Ann Arbor Science, 1975, P78)

Chapter 4
Air Pollution

The air we breathe, like the water we drink, the food we eat, is necessary to our life. And as with water and food, we want to be assured that the air will not cause any harm. We expect to breathe "clean air". However, because of human activities, large amounts of toxic, harmful substances are discharged into the air, which makes the air quality deteriorate sharply and adversely affects human health. So the man started to study how to control air pollution.

This chapter first introduces some basic concepts and the sources and effects of some major air pollutants. Then some of the alternatives available for treating emissions are discussed. Finally, some basic meteorology, which illustrates how the transport and dispersion of pollutants take place, is discussed, followed by an introduction of the methods of predicting air pollutant concentrations.

4.1 GET TO KNOW AIR POLLUTION

4.1.1 Air Pollution

Modern atmosphere is mainly composed of dry clean air, water vapor and suspended particles.

Dry clean air is composed of a variety of ingredients, which can be roughly divided into two categories: constant composition and variable component. The percentage of constant composition, such as nitrogen, oxygen, argon, helium, neon, krypton, and xenon, basically does not change in a different time and place, that is, it is relatively fixed in the atmosphere. On the other hand, the content of variable component, such as carbon dioxide, methane, nitrogen oxides, sulfur oxides, and ozone, whose formation is closely related with human activities, obviously changes in a different time and place. The content of variable component in the atmosphere is far less than that of constant component, but their influence on air quality is more remarkable.

Water vapor in the atmosphere plays an important role in climate change, and its content in the atmosphere significantly changes with time and place.

Suspended particles are impurities in the atmosphere. They are mainly derived from natural processes, such as rock weathering, volcano eruptions, tsunamis etc., and their content in the atmosphere also varies with time and locations.

Certain pollutants that enter the atmosphere through natural or artificial emissions, cumulate exceeding pollutant concentrations (far more than the natural concentration value), which stay for a certain time may do harm to human health and ecological environment, which refers to ***air pollution***.

One of the causes of air pollution comes from human activities that various pollutants are discharged into the atmosphere in the process of production and human life. Another reason comes from the natural process, such as volcano eruptions, forest-fire, rock weathering etc., which may also release a variety of pollutants into the atmosphere.

The formation process of atmospheric pollution is composed of three parts: the emission source, the atmosphere and the receiver, as in Figure 4.1:

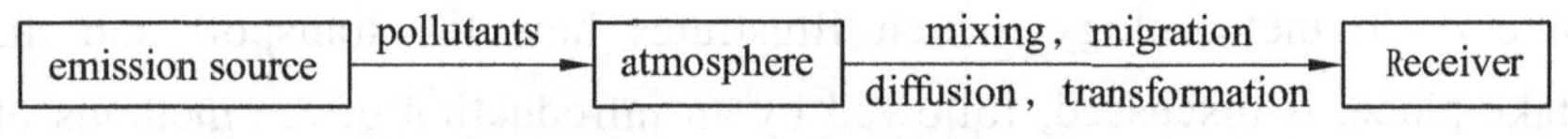

Figure 4.1 The formation process of atmospheric pollution

The pollutants are discharged from the emission source into the atmosphere, through a series of atmospheric motion process, such as mixing, migration, diffusion and chemical transformation, and finally reach the receiver. Lacking of any one aspect will not form air pollution.

Since atmosphere is as the carrier of air pollutants and different atmospheric motions have different scales, so the influence range of air pollution also has different dimensions: local air pollution, regional air pollution, and global air pollution.

4.1.2 Air Pollutants

Air pollutants are substances which, when present in the atmosphere, adversely affect the health of humans, animals, plants, or microbial life; damage materials, or interfere with the enjoyment of life and the use of property.

For the purposes of this discussion, we shall limit our consideration of pollutants to the following seven pollutants.

Particulate Pollutants

Particulate pollutants refer to the dispersoid particles in the form of solid or liquid exist in the air. They are classified as follows:

Dust Dust is solid particles that are 1) entrained by process gases directly from the material being handled or processed, e.g., coal, ash, and cement; 2) direct offspring of a parent material undergoing a mechanical operation, e.g., saw dust from woodworking; 3) entrained materials used in a mechanical operation, e.g., sand from sandblasting. Dust consists of relatively large particles. Cement dust, for example, is about 100 μm in diameter.

Fume Fume is a solid particle, frequently a metallic oxide, formed by the condensation of vapors by sublimation, distillation, calcination, or chemical reaction processes. Examples of fumes are zinc and lead oxide resulting from the condensation and oxidation of metal volatilized in a high-temperature process. The particles in fumes are quite small, with diameters from 0.03 to 0.3 μm.

Mist Mist is a liquid particle formed by the condensation of a vapor and perhaps by chemical reaction. An illustration of this process is the formation of sulfuric acid mist,

$$SO_3(\text{gas})\ 22\ ^\circ C \longrightarrow SO_3\ (\text{liquid}) \quad (4\text{-}1)$$

$$SO_3\ (\text{liquid}) + H_2O \longrightarrow H_2SO_4 \quad (4\text{-}2)$$

Sulfur trioxide gas becomes a liquid since its dew point is 22 °C, and SO_3 particles are hydroscopic. Mists typically range from 0.5 to 3.0 μm in diameter.

Smoke Smoke is solid particles formed as a result of incomplete combustion of carbonaceous materials. Although hydrocarbons, organic acids, sulfur oxides, and nitrogen oxides are also produced in combustion processes, only the solid particles resulting from the incomplete combustion of carbonaceous materials are smoke. Smoke particles have diameters from 0.05 to approximately 1 μm.

Spray Spray is a liquid particle formed by the atomization of a parent liquid.

Sulfur oxides

Sulfur oxides mainly refer to sulfur dioxide (SO_2), hydrogen sulfide (H_2S), and sulfate salts. Sulfur oxides may be both primary and secondary pollutants. Power plants, industry, volcanoes, and the oceans emit SO_2, SO_3, and SO_4^{2-} directly as primary pollutants. In addition, biological decay processes and some industrial sources emit H_2S, which is oxidized to form the secondary pollutant SO_2.

The most important oxidizing reaction for H_2S appears to be one involving ozone:

$$H_2S + O_3 \longrightarrow H_2O + SO_2 \quad (4\text{-}3)$$

The combustion of fossil fuels containing sulfur yields sulfur dioxide in direct proportion to the sulfur content of the fuel:

$$S + O_2 \longrightarrow SO_2 \quad (4\text{-}4)$$

This reaction implies that for every gram of sulfur in the fuel, two grams of SO_2 are emitted to the atmosphere. Because the combustion process is not 100% efficient, we generally assume that 5% of the sulfur in the fuel ends up in the ash, that is, 1.90 g SO_2 per gram of sulfur in the fuel is emitted.

The ultimate fate of most of the SO_2 in the atmosphere is conversion to sulfate salts, which are removed by sedimentation or by washout with precipitation. The conversion to sulfate is by either of two routes: catalytic oxidation or photochemical oxidation. The first process is most effective if water droplets containing Fe^{3+} or Mn^{2+} or NH_3 are present:

$$2SO_2 + 2H_2O + O_2 \xrightarrow{\text{catalyst}} 2H_2SO_4 \quad (4\text{-}5)$$

At low relative humilities, the primary conversion process is photochemical oxidation. The first step is photoexcitation of the SO_2:

$$SO_2 + h\nu \longrightarrow \dot{S}O_2 \quad (4\text{-}6)$$

The excited molecule then readily reacts with O_2 to form SO_3:

$$\dot{S}O_2 + O_2 \longrightarrow SO_3 + O \quad (4\text{-}7)$$

The trioxide is very hygroscopic and consequently is rapidly converted to sulfuric acid:

$$SO_3 + H_2O \longrightarrow H_2SO_4 \quad (4\text{-}8)$$

This reaction in a large part accounts for acid rain (that is, precipitation with a pH value less than 5.6) found in industrialized areas. Normal precipitation has a pH of 5.6 due to the carbonate buffer system.

Nitrogen Oxides

Nitrogen oxides mainly refer to nitrogen monoxide (NO), nitrogen dioxide (NO_2), and secondary pollutants. Bacterial action in the soil releases nitrous oxide (N_2O) to the atmosphere. In the upper troposphere and stratosphere, atomic oxygen reacts with the nitrous oxide to form nitric oxide(NO).

$$N_2O + O \longrightarrow 2NO \quad (4\text{-}9)$$

The atomic oxygen results from the dissociation of ozone. The nitric oxide further reacts with ozone to form nitrogen dioxide (NO_2).

$$NO + O_3 \longrightarrow NO_2 + O_2 \tag{4-10}$$

Combustion processes account for 96 percent of the anthropogenic sources of nitrogen oxides. Although nitrogen and oxygen coexist in our atmosphere without reaction, their relationship is much less indifferent at high temperatures and pressures. At temperatures in excess of 1,600 K, they react.

$$N_2 + O_2 \rightleftharpoons 2NO \tag{4-11}$$

If the combustion gas is rapidly cooled after the reaction by exhausting it to the atmosphcrc, the reaction is quenched and NO is the by-product. The NO in turn reacts with ozone or oxygen to form NO_2.

Ultimately, the NO_2 is converted to either NO_2^- or NO_3^- in particulate form. The particulates are then washed out by precipitation. The dissolution of nitrate in a water droplet allows for the formation of nitric acid (HNO_3). This, in part, accounts for "acid" rain found downwind of industrialized areas.

Carbon monoxide

Incomplete oxidation of carbon results in the production of carbon monoxide. The natural anaerobic decomposition of carbonaceous material by soil microorganisms releases methane (CH_4) to the atmosphere each year on a world-wide basis. The natural formation of CO results from an intermediate step in the oxidation of the methane.

Anthropogenic sources (those associated with the activities of human beings) include motor vehicles, fossil fuel burning for electricity and heat, industrial processes, solid waste disposal, and miscellaneous burning of such things as leaves and brush. Motor vehicles account for more than 60 percent of the emission.

No significant change in the global atmospheric CO level has been observed over the past years. Yet the worldwide anthropogenic contribution of combustion sources has doubled over the same time period. Since there is no apparent change in the atmospheric concentration, a number of mechanism (sinks) have been proposed to account for the missing CO. The two most probable are: 1) Reaction with hydroxyl radicals to form carbon dioxide; 2) Removal by soil microorganisms. It has been estimated that these two sinks annually consume an amount of CO that just equals the production.

Hydrocarbons

Hydrocarbons that exist in the gas phase are generally considered to be those with five or fewer carbon atoms. Methane and terpenes are the two major hydrocarbons emitted by natural sources. Unlike methane, which results from decaying organic matter, terpenes are released from living plants. Terpenes are unsaturated hydrocarbons. They are the compounds that give the characteristic scent to lemon and pine.

The major anthropogenic sources are partially burned gasoline and incinerator emissions. Many of the hydrocarbons are oxidized. Several of the hydrocarbons are converted to other organic compounds in the presence of nitrogen oxides. Ultimately they may be converted to particles.

Lead

Volcanic activity and airborne soil are the primary natural sources of atmospheric lead. Smelters and refining processes, as well as incineration of lead-containing wastes, are major point sources of lead. Approximately 70 to 80 percent of the lead added to gasoline is eventually discharged to the atmosphere.

Submicron lead particles, which are formed by volatilization and subsequent condensation, attach to larger particles or form nuclei before they are removed from the atmosphere. Once they have attained a size of several microns, they either settle out or are washed out by rain.

Photochemical oxidants

Unlike other pollutants, the photochemical oxidants result entirely from atmospheric reactions and are not directly attributable to either people or nature. Thus, they are called ***secondary pollutants***. They are formed through a series of reactions that are initiated by the absorption of a photon by an atom, molecule, free radical, or ion.

Ozone is the principal photochemical oxidant. Its formation is usually attributed to the nitrogen dioxide photolytic cycle. Hydrocarbons modify this cycle by reacting with atomic oxygen to form free radicals (highly reactive organic species). The hydrocarbons, nitrogen oxides, and ozone react and interact to produce more nitrogen dioxide and ozone.

The whole reaction sequence is dependent upon an abundance of sunshine. In the lowest thousand meters of the atmosphere, ozone and other compounds are generated as secondary pollutants as the result of atmospheric reactions. A result of these reactions is the photochemical "smog" for which Los Angeles is famous.

4.1.3 Units of Air Pollutant concentrations

There are several ways to express the concentration of air pollutants. The two basic units of measure used in reporting air pollution data are micrograms per cubic meter ($\mu g/m^3$) and parts per million (ppm). Both $\mu g/m^3$ and ppm are used to indicate the concentration of a gaseous pollutant. However, the concentration of particulate matter may be reported only as $\mu g/m^3$.

The conversion between $\mu g/m^3$ and ppm is based on the fact that at standard conditions (0 °C and 101.325 kPa), one mole of an ideal gas occupies 22.414 L. Thus, we may write an equation that converts the mass of the pollutant M_p in grams to its equivalent volume V_p in liters at standard temperature and pressure (STP):

$$V_p = \frac{M_p}{\text{GMW}} \times 22.414 \ \text{L/GM} \tag{4-12}$$

where GMW is the gram molecular weight of the pollutant. For readings made at temperatures and pressures other than standard conditions, the standard volume, 22.414 L/GM, must be corrected. We use the ideal gas law to make the correction:

$$22.414 \ \text{L/GM} \times \frac{T_2}{273 \ \text{K}} \times \frac{101.325 \ \text{kPa}}{P_2} \tag{4-13}$$

where t_2 and p_2 are the absolute temperature and absolute pressure at which the readings were made. Since ppm is a volume ratio, we may write

$$\text{ppm} = \frac{V_p}{V_a} \tag{4-14}$$

where V_a is the volume of air in cubic meters at the temperature and pressure of the reading. We than combine Equations 4-12, 4-13, and 4-14 to form Equation 4-15.

$$\text{ppm} = \frac{\frac{M_p}{\text{GMW}} \times 22.414 \times \frac{T_2}{273 \ \text{K}} \times \frac{101.325 \ \text{kPa}}{P_2}}{V_a \times 1{,}000 \ \text{L/m}^3} \tag{4-15}$$

where M_p is in μg. The factors converting μg to g and L to millions of L cancel one another. Unless otherwise stated, it is assumed that $V_a = 1.00 \ m^3$.

Example 4.1 Converting $\mu g/m^3$ ppm

A one-cubic-meter sample of air was found to contain 80 $\mu g/m^3$ of SO_2. The temperature and pressure were 25 °C and 103.193 kPa when the air sample was taken. What was the SO_2 concentration in ppm?

Solution First we must determine the GMW of SO_2.

$$\text{GMW of } SO_2 = 32.07 + 2 \times (16.00) = 64.07$$

Next we must convert the temperature to absolute temperature. Thus,

$$25 \ °\text{C} + 273 \ \text{K} = 298 \ \text{K}$$

Now we may make use of Equation 4-15.

$$\text{ppm} = \frac{\frac{80}{64.07} \times 22.414 \times \frac{298}{273} \times \frac{101.325}{103.193}}{1.00 \times 1{,}000} = 0.0300 \ \text{ppm of } SO_2$$

4.1.4 Sources of Air Pollution

The sources of air pollution can be categorized as two types: natural sources of air pollution and people-made sources of air pollution.

Many of the pollutants of concern are formed and emitted through natural processes. For example, naturally occurring particulates include pollen grains, fungus spores, salt spray, smoke particles from forest fires, and dust from volcanic eruptions. Gaseous pollutants from natural sources include carbon monoxide as a product of animal and plant respiration, hydrocarbons from conifers, sulfur dioxide from geysers, hydrogen sulfide from the breakdown of cysteine and other sulfur-containing amino acids by bacterial action, nitrogen oxides, and methane.

Near cities and in populated areas, more than 90% of the volume of air pollutants is the result of human activity. People-made sources of air pollution may be broadly classified as stationary and mobile sources. Stationary sources include combustion processes, solid waste disposal, industrial processes, and construction and demolition. The principal pollutants from stationary combustion processes are particulate matter, in the form of fly ash and smoke, and the gaseous pollutants sulfur dioxide, nitric oxide, and nitrogen dioxide.

4.1.5 Effects of Air Pollution

The effects of air pollution are considerable and air pollution will impact many aspects of our environment: visually aesthetic resources, plants and animals, soils, water quality, materials, and human health. Air pollutants affect visual resources by discoloring the atmosphere, reducing visual range and atmospheric clarity, so that the visual contrast of distant objects is decreased. We can't see as far in polluted air, and what we do see has less color contrast. Reduction in visibility results in a social cost due to slowdown of air traffic and the need for instrument-guided landing systems.

In the air pollution episodes, the local presence of gases was objectionable because of odor, taste, or obvious corrosive or chemical effects. Today these gross sensory insults are rarely encountered. However subtle health effects persist, such as eye or nose irritation or difficult breathing. In the extreme, health effects extend to the brain (CO) and stomach (several pollutants alone or in combination). Damage to vegetation due to chronic exposure to atmospheric pollutants may be one of the more apparent precursor symptoms leading to identification of chronic air pollution.

Effects on Human Health

Health effects were the dominant considerations in early air pollution episodes for obvious reasons. While the specific pollutant or groups of pollutants generating the observed effects frequently could not be identified, there was sufficient information to implicate certain pollutants as significant contributors. Early research to relate pollutant concentrations and effects was concentrated on these clearly identifiable pollutants.

The human respiratory system is quite efficient in filtering the larger particles out of the air we breathe. Particles smaller than about 5 μm, however, can penetrate to the lungs and be deposited in the alveoli. For example, cigarette smoke particles are smaller than 1 μm, and they enter the lungs and are deposited in the alveoli.

Some particles are particularly damaging because they adsorb gases which cause more intense irritation locally. Gases also penetrate into the deepest lung pockets. Both particles and gases entering the body through the respiratory system can affect the gastrointestinal system. Some chemicals, such as lead, can enter the human bloodstream either from the digestive system (ingestion) or by passing through the lung membranes (the respiratory system), and airborne tritium and a few other chemicals can enter the bloodstream through the skin.

Each pollutant affects the human body differently. The colorless and odorless CO is lethal to humans within a few minutes at concentrations exceeding 5,000 ppm. CO reacts with hemoglobin in the blood to form carboxyhemoglobin (COHb). Hemoglobin has a greater affinity for CO than it does for oxygen. Thus, the formation of COHb effectively deprives the body of oxygen. At COHb levels of 5 to 10 percent, visual perception, manual dexterity, and ability to learn are impaired. A concentration of 5 ppm of CO for eight hours will result in a COHb level of about 7.5 percent. At COHb levels of 2.5 to 3 percent, people with heart disease are not able to perform certain exercises as well as they might be in the absence of COHb. The sensitive populations are those with heart and circulatory ailments, chronic pulmonary disease, developing fetuses, and those with conditions that cause increased oxygen demand, such as fever from an infection disease.

Exposure to NO_2 concentrations above 5 ppm for 15 minutes results in cough and irritation of the respiratory tract. Continued exposure may produce an abnormal accumulation of fluid in the lung. The gas is reddish brown in concentrated form and gives a brownish yellow tint at lower concentrations.

The sulfur oxides include sulfur dioxide (SO_2), sulfur trioxide (SO_3), their acids, and the salts of their acids. There is speculation that a definite synergism exists whereby fine particulates carry absorbed SO_2 to the lower respiratory tract (LRT). The SO_2 in the absence of particulates would be absorbed in the mucous membranes of the upper respiratory tract (URT).

Effects on Plants and Animals

The detrimental impacts of air pollution are not limited to those involving human health. Plants and animals are also susceptible. For example, fluorine is emitted from aluminum, glass, phosphate, fertilizer, and some clay-baking operations in significant quantities. Frequently, the plant damage is observed on the fruit or on the flowers, either of which significantly lowers the value of the crop. Fluorine affects plants at concentrations several orders of magnitude below that at which human health is affected.

Fluorine also has an effect at even lower concentrations when it is taken up by shrubs, trees, or grass which are subsequently eaten by cattle or other animals. The animals may develop fluorosis, although the plants may not show signs of damage. The animals act as concentrators of the fluorine, resulting in poor health and associated lower animal value or survival capability. Farm animals, particularly cattle, sheep, and swine, are susceptible to fluorine poisoning. Fluorosis is characterized by mottled teeth and a condition of the joints known as exostosis leading to lameness and ultimately death. Some heavy metals, such as mercury and lead, and most radionuclides also become concentrated in plants and animals, frequently in specific organs.

Different plants and animals have different susceptibilities to pollutants. For example, sugar maple can tolerate relatively high concentrations of sulfur dioxide alone, but it is susceptible to damage from exposure to SO_2 and O_3 together. White pine, on the other hand, is very sensitive to damage from either pollutant alone.

Effects on Materials

Perhaps the most familiar effect of air pollution on materials is soiling of building surfaces, clothing, and other articles. Soiling results from the deposition of smoke on surfaces. Over time this deposition becomes noticeable as soiling, a discoloring or darkening of the surface. Damage to the surface , of course, results from the cleaning operation. In the case of exterior building materials, sandblasting is often required to clean the surface, and part of the surface is removed in the process of cleaning.

Another effect of air pollution is that of accelerating the corrosion of metals. Sulfur and nitrogen oxides react in the atmosphere to form acidic compounds which attack metal surfaces, a problem which has been particularly acute for the communications, switchgear, and computer industries.

One of the early noted effects of the Los Angeles smog was rubber cracking. Indeed, the effect of ozone, a principle ingredient of smog, on rubber is so specific that rubber cracking may be used to measure ozone concentrations. Fluorine is particularly reactive, and at high atmospheric concentrations etching of glass has been observed. These impacts are taken into consideration when sensitive components are designed, and the required protective measures or design modifications add to the cost of the item being produced.

Fabrics are also affected by air pollutants. They can be disintegrated by a fine sulfuric acid mist present in the air. Other effects of pollutants on fabrics include bleaching and discoloration, both of which lead to an unacceptable deterioration of products made of fabrics.

The action of hydrogen sulfide on lead-base paints is well known. Hydrogen sulfide, in the presence of moisture, reacts with lead oxide in white paint to form lead sulfide, producing a brown to black discoloration. As might be imagined, this is unattractive on a white house.

4.2 CONTROL OF AIR POLLUTION

4.2.1 Ambient Air Quality Standards

Ambient air quality standards are important because they are tied to emission standards that attempt to control air pollution. Many countries have developed ambient air quality standards.

Ambient Air Quality Standards in U.S.

National Ambient Air Quality Standards (NAAQS) for the United States define two levels or types of air quality standards. Primary standards are levels that are set to protect the health of people, but may not protect against damaging effects of air pollution to structures, paint, and plants. Secondary standards are designed to protect from other environmental degradation resulting from air pollution; however, most secondary levels are the same , or nearly so, as the primary levels.

The Environmental Protection Agency （EPA） Office of Air Quality Planning and Standards (OAQPS) has set National Ambient Air Quality Standards for six principal pollutants, which are called "criteria" pollutants. They are listed in Table 4.1. Units of measure for the standards are parts per million (ppm) by volume, parts per billion (ppb – 1 part in 1,000,000,000) by volume, milligrams per cubic meter of air (mg/m^3), and micrograms per cubic meter of air ($\mu g/m^3$).

Table 4.1 National ambient air quality standards in U.S.

	Primary Standards		Secondary Standards	
Pollutant	Level	Averaging Time	Level	Averaging Time
Carbon Monoxide	9 ppm (10 mg/m^3)	8-hour	None	
	35 ppm (40 mg/m^3)	1-hour		
Lead	0.15 $\mu g/m^3$	Rolling 3-Month Average	Same as Primary	
Nitrogen Dioxide	53 ppb	Annual (Arithmetic Average)	Same as Primary	
	100 ppb	1-hour	None	
Particulate Matter (PM_{10})	150 $\mu g/m^3$	24-hour	Same as Primary	
Particulate Matter (PM2.5)	15.0 $\mu g/m^3$	Annual (Arithmetic Average)	Same as Primary	
	35 $\mu g/m^3$	24-hour	Same as Primary	
Ozone	0.075 ppm (2008 std)	8-hour	Same as Primary	
	0.08 ppm (1997 std)	8-hour	Same as Primary	
	0.12 ppm	1-hour	Same as Primary	
Sulfur Dioxide	0.03 ppm (1971 std)	Annual (Arithmetic Average)	0.5 ppm	3-hour
	0.14 ppm(1971 std)	24-hour		
	75 ppb	1-hour	None	

Ambient Air Quality Standards in China

National ambient air quality standards for China are divided into three function areas. A class of region implies nature reserves, scenic areas and other areas which need special protection; B class of region implies the urban planning residential areas, commercial and traffic as well as residential areas, cultural areas, general industrial area and rural areas; C class of region implies special industrial park. At the same time, ambient air quality standards are divided into three grades. A region executes Ⅰ grade standard, B region executes Ⅱ grade standard, and C region executes Ⅲ grade standard. The specific standard values are shown in Table 4.2.

Table 4.2 National ambient air quality standards in China

<table>
<tr><th rowspan="2">Pollutants</th><th rowspan="2">Sampling Time</th><th colspan="6">Concentration Limited Values</th><th rowspan="2">Units</th></tr>
<tr><th colspan="2">Grade Ⅰ</th><th colspan="2">Grade Ⅱ</th><th colspan="2">Grade Ⅲ</th></tr>
<tr><td rowspan="3">SO_2</td><td>Annual Average</td><td colspan="2">0.02</td><td colspan="2">0.06</td><td colspan="2">0.10</td><td rowspan="16">mg/m^3 (standard state)</td></tr>
<tr><td>Day Average</td><td colspan="2">0.05</td><td colspan="2">0.15</td><td colspan="2">0.25</td></tr>
<tr><td>1-hour Average</td><td colspan="2">0.15</td><td colspan="2">0.50</td><td colspan="2">0.70</td></tr>
<tr><td rowspan="2">TSP</td><td>Annual Average</td><td colspan="2">0.08</td><td colspan="2">0.20</td><td colspan="2">0.30</td></tr>
<tr><td>Day Average</td><td colspan="2">0.12</td><td colspan="2">0.30</td><td colspan="2">0.50</td></tr>
<tr><td rowspan="2">PM_{10}</td><td>Annual Average</td><td colspan="2">0.04</td><td colspan="2">0.10</td><td colspan="2">0.15</td></tr>
<tr><td>Day Average</td><td colspan="2">0.05</td><td colspan="2">0.15</td><td colspan="2">0.25</td></tr>
<tr><td rowspan="3">NO_x</td><td>Annual Average</td><td colspan="2">0.05</td><td colspan="2">0.05</td><td colspan="2">0.10</td></tr>
<tr><td>Day Average</td><td colspan="2">0.10</td><td colspan="2">0.10</td><td colspan="2">0.15</td></tr>
<tr><td>1-hour Average</td><td colspan="2">0.15</td><td colspan="2">0.15</td><td colspan="2">0.30</td></tr>
<tr><td rowspan="3">NO_2</td><td>Annual Average</td><td colspan="2">0.04</td><td colspan="2">0.04</td><td colspan="2">0.08</td></tr>
<tr><td>Day Average</td><td colspan="2">0.08</td><td colspan="2">0.08</td><td colspan="2">0.12</td></tr>
<tr><td>1-hour Average</td><td colspan="2">0.12</td><td colspan="2">0.12</td><td colspan="2">0.24</td></tr>
<tr><td rowspan="2">CO</td><td>Day Average</td><td colspan="2">4.00</td><td colspan="2">4.00</td><td colspan="2">6.00</td></tr>
<tr><td>1-hour Average</td><td colspan="2">10.00</td><td colspan="2">10.00</td><td colspan="2">20.00</td></tr>
<tr><td>O_3</td><td>1-hour Average</td><td colspan="2">0.12</td><td colspan="2">0.16</td><td colspan="2">0.20</td></tr>
<tr><td rowspan="2">Pb</td><td>Quarter Average</td><td colspan="6">1.50</td><td rowspan="5">$\mu g/m^3$ (standard state)</td></tr>
<tr><td>Annual Average</td><td colspan="6">1.00</td></tr>
<tr><td>B[α]P</td><td>Day Average</td><td colspan="6">0.01</td></tr>
<tr><td rowspan="4">F</td><td>Day Average</td><td colspan="6">7[1]</td></tr>
<tr><td>1-hour Average</td><td colspan="6">20[1]</td></tr>
<tr><td>Month Average</td><td colspan="3">1.8[2]</td><td colspan="3">3.0[3]</td><td rowspan="2">$\mu g/(dm^2 \cdot d)$</td></tr>
<tr><td>Plant Growth Quarter Average</td><td colspan="3">1.2[2]</td><td colspan="3">2.0[3]</td></tr>
</table>

① adapted for urban region;

② adapted for animal husbandry region and animal husbandry with farming region(以牧业为主的半农半牧区), silkworm region;

③ adapted for agriculture and forestry regions.

4.2.2 Air Quality Control

The objective of air pollution control is to maintain an atmosphere in which pollutants have no negative impact on human activities, materials, and plants. Obviously, the best way to control air pollution is to eliminate the sources of the pollution. For example, lead emissions from automobiles are eliminated by burning nonleaded fuels, and nitrogen oxide emissions have been significantly reduced by redesigning engines. We do not know how to eliminate nitrogen oxide emissions completely, but changing our means of transportation might help. A possible alternative is to shift the location of the nitrogen oxides emissions, for example, from the automobile tailpipe to the stack of all electric generating station, by using electric or hydrogen-fueled cars. Legislating the quantity of ash and sulfur in fuels is a way of reducing emissions of these materials or their end products.

Other solutions include reducing emissions by using "add-on" devices. In the case of the automobile, carbon canisters are used to adsorb hydrocarbon vapors emitted from the carburetor and the gas tank. The vapors are subsequently returned to the engine for burning. In the automobile exhaust system, catalytic converters chemically reduce emissions of hydrocarbons. The energy in these hydrocarbons is lost as heat. In industrial applications, scrubbers (absorbers) may be used to remove pollutants from gas streams and possibly to induce them to react chemically to form more stable substances for collection and storage.

Planned dispersion may be used to control local air quality where emissions are not controllable by other techniques. Emissions from tall stacks have more time to disperse in the atmosphere before reaching the ground where they impact on humans, materials, and vegetation.

Various air pollution control devices are conveniently divided into those applicable for controlling particulates and those used for controlling gaseous pollutants. The reason, of course, is the difference in sizes.

4.2.3 Particle Emission Control

Designers of particle emission control equipment must deal with solid and liquid particles ranging from smaller than 1 μm to larger than 100 μm in diameter. The smaller particles are far more difficult to collect. Collectors are broadly categorized according to the physics of the collecting mechanism.

According to the mechanism of dust removal, the devices for controlling particulates can be divided into four types: mechanical dust removal devices, wet collectors, electrostatic precipitators, and fabric and fibrous mat collectors.

Mechanical Dust Removal Devices

Gravitational settling chambers Gravitational settling chambers are simple, inexpensive collectors in which gravitational forces dominate vertical particle motions. They are essentially simple expansions in a duct in which the horizontal velocity of the particles is reduced to allow time for the particles to settle out by gravity.

Figure 4.2 is a sketch of a very simple gravitational settling chamber. A theoretical expression for the efficiency of this collector is

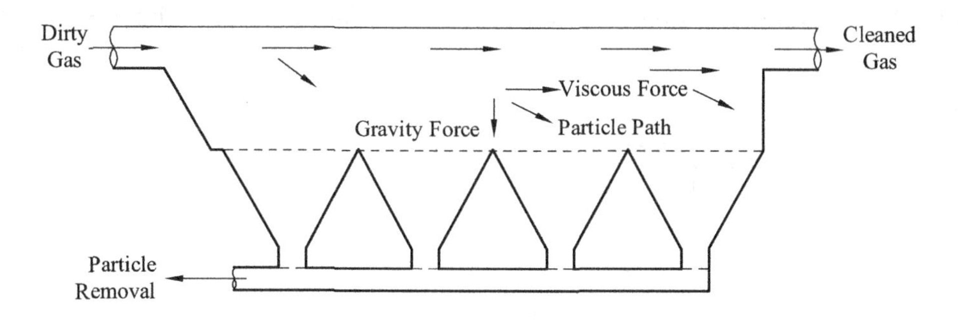

Figure 4.2 Gravitational settling chamber

(Source: J. Glynn Henry and Gary W. Heinke, *Environmental Science And Engineering*, Prentice Hall, 1989, P499)

$$\eta_{\mathrm{g}} = 1 - \exp\left\{ -\frac{g d_{\mathrm{p}}^{2} \rho_{\mathrm{p}} L}{18\, \mu u H} \right\} \tag{4-16}$$

where $\exp\{A\}$ is used to represent the exponential e^{A}, e being the base for natural logarithms, and where

η_{g} = efficiency of removal, as a fraction

L = length of the collector in m

H = depth of the collector in m

u = horizontal velocity of the gas and particles through the collector in m/s

ρ_{p} = density of the particle in kg/m^3

D_{p} = diameter of the particle in m

μ = dynamic viscosity of air in kg/m · s

Example 4.2

Calculate the 50 percent cutoff diameter for particles of CaO suspended in an airstream at 100 °C and at atmospheric pressure for a gravitational settling chamber 3 m long and 1 m high when the gas velocity in the collector is 1 m/s. The 50 percent cutoff diameter is defined as the particle diameter at which $\eta_g = 50$ percent, i.e., 50 percent of the particles of this diameter are collected and 50 percent are lost.

Solution Using Equation 4-16,we obtain

$$\eta_g = 0.5 = 1 - \exp\left\{-\frac{g d_p^2 \rho_p L}{18\ \mu u H}\right\}$$

ρ_p for CaO=3,310 kg/m^3, and μ for air=2.17 × 10^{-5} N · s/m^2=2.17 × 10^{-5} kg/m · s(at 100 °C).

Therefore,

$$0.5 = \exp\left\{-\frac{9.81 \times 3\ 310 \times 3 \times d_p^2}{18 \times 2.17 \times 10^{-5}}\right\}$$

so that

$$\begin{aligned} d_p^2 &= \ln(0.5) \times \left\{-\frac{18 \times 2.17 \times 10^{-5}}{9.81 \times 3\,310 \times 3}\right\} \\ &= (-0.693) \times (-40 \times 10^{-10}) \\ &= 28 \times 10^{-10}\ \text{m}^2 \end{aligned}$$

Therefore, the 50 percent cutoff particle diameter $d_p/50$=5.3 × 10^{-5} m=53 μm

Inertial collectors A very simple particle skimmer is sketched in Figure 4.3(a), and a common particle separator called a cyclone is sketched in Figure 4.3(b).

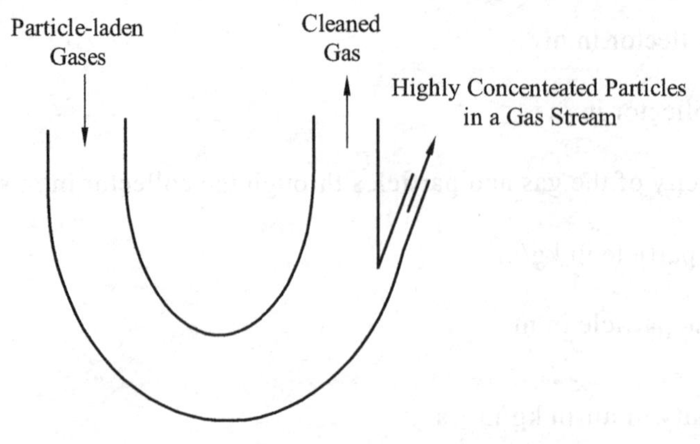

(a) A very simple centrifugal particle skimmer

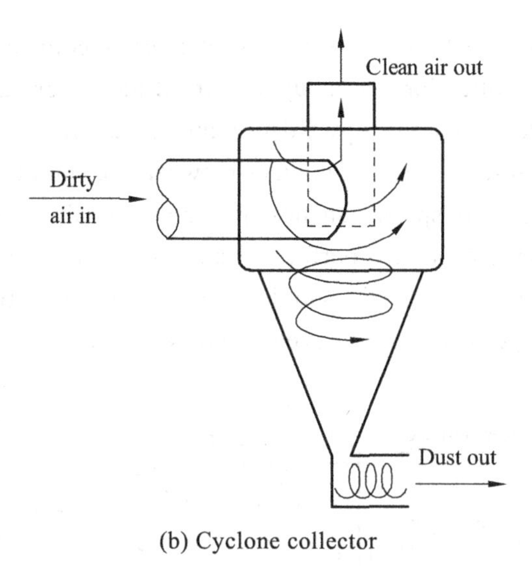

(b) Cyclone collector

Figure 4.3 Simple inertial particle collectors

(Source: J. Glynn Henry and Gary W. Heinke, *Environmental Science And Engineering*, Prentice Hall, 1989, P501)

Both collectors rely on centrifugal forces to separate the heavier particles from the lighter gas molecules. The skimmer shown in Figure 4.3(a) simply increases the particle concentration in a separated gas stream, which might then be passed through a gravitational collector possibly through a cyclone like the one in Figure 4.3(b).

As shown in Figure 4.3(b), particle-laden gases enter the cyclone at the top and, theoretically, spiral downward along the casing in solid body rotation. Particles migrate to the outside of the spiral, where they slide down the casing to the hopper bottom. The only exit for gases from the cyclone is vertically upward through the central pipe, and to exit, the spiral must contract to a smaller diameter. The decrease in the radius of the particles' trajectory results in increased centrifugal force as the particles move toward the inner spiral.

A typical collection efficiency for a cyclone 1 m in diameter might be 50 percent for 20 μm particles, and a high-efficiency cyclone might have a collection efficiency η_c=80 percent for d_p>10 μm. The typical pressure drop through a conventional cyclone is 5 to 15 cm of water, and through a high-efficiency cyclone is 10 to 30 cm of water.

Gravitational settling chambers and simple inertial separators contain no moving parts. They may be fabricated using metals that can withstand high temperatures and resist corrosive attack by particles or gases. They are equally effective for solid or liquid particles.

Wet Collectors

Wet collectors, or scrubbers, are designed to increase particle sizes using water or slurry droplets, because larger particles are easier to collect. There are many different scrubber designs, but we shall consider only two types here: conventional and venturi scrubbers.

Conventional scrubbers Figure 4.4 is a sketch of a conventional scrubber showing several modes of particle collection. In the upper part of the tower, falling water drops collide with and collect particles from the upward-flowing gases. In the packed section, special shapes are added to increase the area of contact between the liquid and the aerosol (gas plus particles). Despite this use of special shapes, plugging remains a problem in this section. Below the packed section is a flooded perforated disc which may support several centimeters of water, allowing contact between the bubbles containing the particles and the liquid. The liquid drains through the perforations to develop another falling-drop collector section.

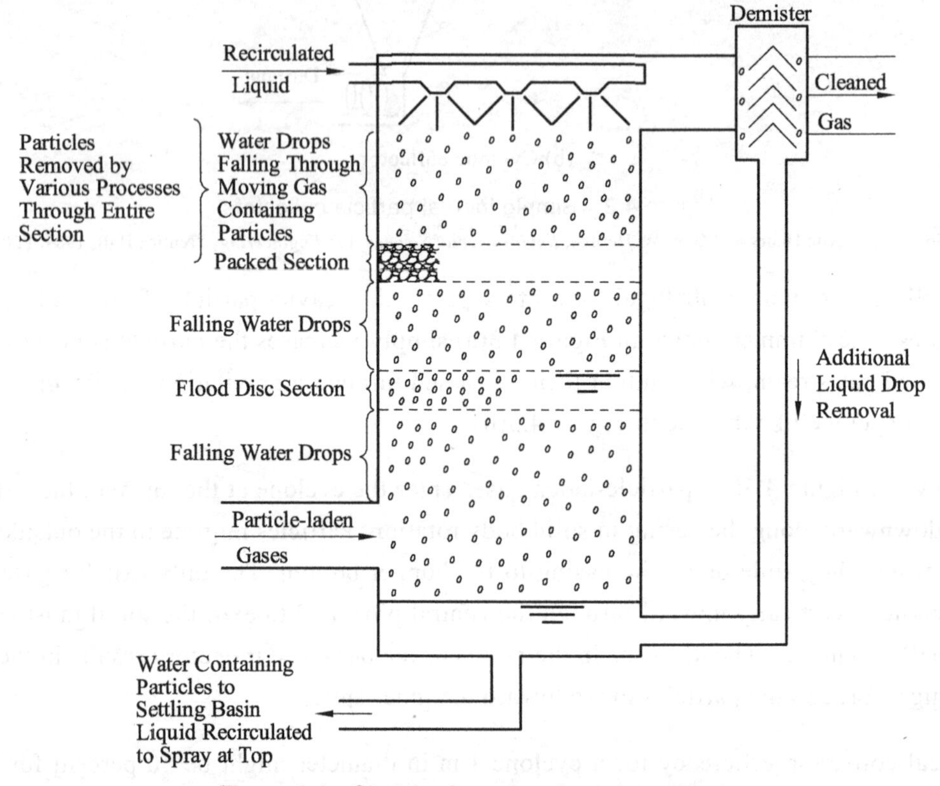

Figure 4.4 Sketch of an absorber or scrubber

(Source: J. Glynn Henry and Gary W. Heinke, *Environmental Science And Engineering*, Prentice Hall, 1989, P503)

Not every collision of a water droplet and a particle results in collection, because of the surface tension of the droplets and particle wettability characteristics. Chemicals are sometimes introduced to reduce the surface tension of the droplet or to improve the ability of the droplets to absorb gases selectively in addition to collecting particles.

The liquid containing the particles is collected at the bottom of the tower and pumped to a settling basin or filtration device, where the particles are removed. The liquid may be recirculated with or without chemical treatment, resulting in a zero-discharge system and reduced makeup water requirements.

The demister at the exit of the scrubber is actually a particle collector. It is designed to remove drops of liquid carried over in the gas stream that leaves the scrubber (called an absorber when the scrubber is designed primarily to control gaseous emissions rather than particle emissions).

The size of the water drops is critical in determining the performance of a scrubber. If the water drops are largely relative to the particles, aerodynamic forces (drag) will displace the particles out of the path of the falling drops, and the number of collisions will decrease significantly. If the drops are the same size as the particles, collisions are also decreased because drag causes the collecting liquid droplets to move with the gas stream and particles. To optimize scrubber performance, the water drops should be a little larger than or smaller than the particles to be collected. The scrubber must be maintained to ensure that the desired drop distribution persists. Scrubber performance is also highly dependent upon the physical and chemical characteristics of the particles, the collecting liquid, and the final droplet (particle) collector.

Venturi scrubbers In venturi scrubbers (Figure 4.5), gases and particles accelerate in the throat followed by rapid deceleration as expansion of the gas stream occurs. Liquid is injected into the venturi throat, usually in a solid stream perpendicular to the gas flow. The high relative velocity between the gas and the liquid streams results in the liquid stream being torn apart, and drops are formed. Particles moving with the gas stream collide with the liquid as the drops are being formed and become entrained in the drops. Drops which are larger than the particles decelerate more slowly than the particles in the expanding section of the venturi, and additional particle collection occurs through this section. The drops containing the particles are subsequently removed as large particles.

The performance of venturi scrubbers is critically dependent upon the gas stream velocity and the physical and chemical properties of the liquid and particles. Thus the venturi scrubber must be operated at a constant gas flowrate if performance is to be maintained for a given particle size distribution and concentration in the gas stream. To overcome this limitation, venturi has been designed in which the throat area can be changed while the device is operating.

The pressure loss Δp through conventional scrubbers ranges from 15 to 40 cm of water. Collector efficiency increases with the pressure loss and may be as high as 95 percent for $d_p>5$ μm. The pressure drop through venturi scrubbers ranges from about 50 to 200 cm of water. At high pressure drops, venturi scrubbers can collect particles as small as 1 μm at efficiencies approaching 99 percent, although this performance is extremely difficult to maintain continuously.

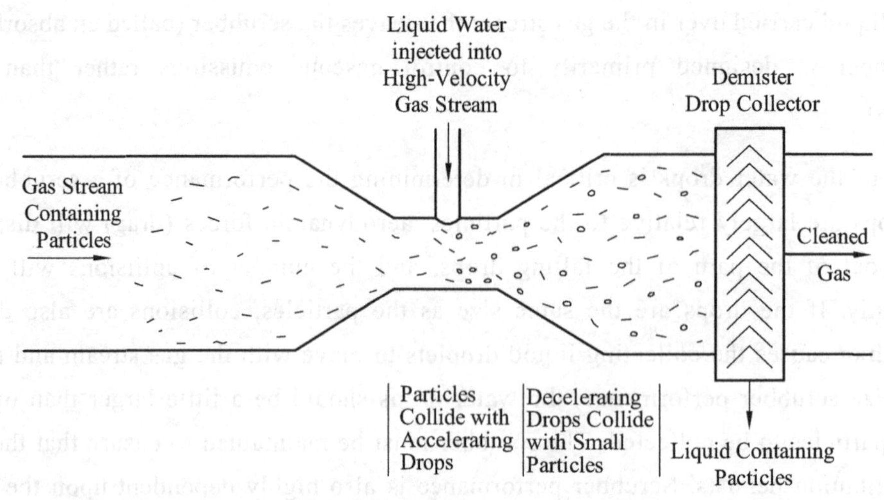

Figure 4.5 Venturi scrubber

(Source: J. Glynn Henry and Gary W. Heinke, *Environmental Science And Engineering*, Prentice Hall, 1989, P504)

Scrubbers collect solid or liquid particles. They can be designed to resist corrosion, and they can be operated at relatively high temperatures as long as the liquid used does not boil and excessive evaporation losses can be prevented. For these reasons, venturi scrubbers are frequently used to collect the small particles generated in steel-making or smelting operations. Operating costs are relatively high for high-pressure loss scrubbers, but capital cost is low compared to other collectors of equivalent performance.

Electrostatic Precipitators

Electrostatic precipitators are widely used in power plants, mainly because power is readily available. Figure 4.6 shows a model of an electrostatic precipitator. The voltage difference between the electrode and the collector plates is maintained at as high a level as possible, but below the field strength at which spark-over occurs. Electrons are released at the electrode in a corona discharge and attach themselves to particles, thus charging the particles. The charged particles or molecules of the same polarity as the electrode migrate toward the grounded surfaces due to electrostatic forces. Positively or negatively charged electrodes may be used. The negative corona generates a slightly greater quantity of O_3 and is slightly more effective for industrial operation.

Migrating ions collide with liquid or solid particles in the gas stream, giving the particles a charge which results in particle motion toward the collector plates.

When particles touch the plates, they stick there. In time, a layer of particles which acts as an insulating blanket will collect on the plates, and the blanket surface charge may actually approach that of the electrode. This blanket must continually be removed by rapping the vertical plates so that the particle layer slides downward, by flooding the plates by washing them with liquid, or, if the particles collected are liquid, by having them run down the plate surfaces much like condensation on a window. The particles falling off the bottom of the plate are collected in hoppers for disposal.

Gas and particle resistivity are important precipitator design variables because they determine the rapidity with which particles can be charged and the field strength in the collector sections. Resistivities vary with temperature and chemical composition. Conditioning agents, i.e., chemicals that significantly change resistivity, such as SO_3 and NH_3, are frequently used to improve collector performance.

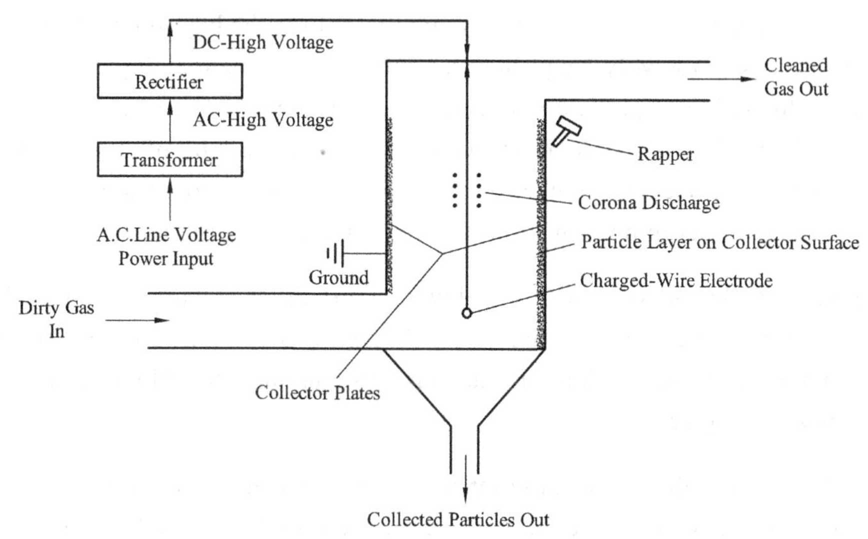

Figure 4.6 An elementary electrostatic precipitator

(Source: J. Glynn Henry and Gary W. Heinke, *Environmental Science And Engineering*, Prentice Hall, 1989, P504)

Precipitator collection efficiencies are as high as 99 percent for particles larger than 2 μm at pressure losses of 5 cm of water or less. The units can be built entirely of metal. They are used almost exclusively in processes discharging corrosive gases at elevated temperatures in very large volumes containing a high percentage of particles larger than 1 μm.

Fire and explosion are always hazards when collecting combustible particles in dry collectors. They are particularly hazardous in electrostatic precipitators because of the danger of ignition by spark-overs.

Fabric and Fibrous Mat Collectors

Fabric or baghouse collectors are designed to remove dry particles from dry, low-temperature (0 to 275 °C) gas stream, which operate like the common vacuum cleaner. Cloth socks 15 to 30 cm in diameter and up to 10 m long are suspended in a chamber, and air forced through the sock, discharges through the fabric. The fabric may be woven or made of felt, but woven cloth is by far the most common. The fabric will remove nearly all particulates, including submicron sizes. Baghouse collectors are widely used in many industrial applications but are sensitive to high temperatures and humidity. A schematic baghouse for control of particulate air pollutants is shown in Figure 4.7.

The cloth from which the bags or socks are made may have holes exceeding 100 μm across, but when properly operated, the collector performs with efficiency greater than 99 percent for particles of d_p>1 μm . Small particles are collected using the filter cake on the cloth surface as the filtration medium. As the thickness of the cake builds up, the pressure loss through the baghouse, and hence power costs, increase.

The filter cake is removed from collectors by simply shaking the bag so that the cake falls off. To avoid the necessity for very frequent bag shaking while maintaining a reasonable filter cake thickness for efficient particle collection without excessive pressure loss, the volume flowrate through the cloth is usually restricted to 0.5 to 2 m^3/s per m^2 of cloth surface. Typical pressure drops across a baghouse range from 5 to 40 cm of water for shaking periods ranging from 4 or 5 times per hour to once in several hours. A typical bag life might be 2 to 3 years.

Fiber mat particle collectors operate at very low pressure drops and are frequently disposable, although many may be washed and reused several times. Fiber mat filters are used extensively in air conditioning and hot-air domestic heating systems, and for filtration of air entering internal combustion engines.

The basic mechanism of dust removal in fabric filters is thought to be similar to the action of sand filters in water treatment. The dust particles adhere to the fabric due to entrapment and surface forces. They are brought into contact by impingment and/or Brownian diffusion.

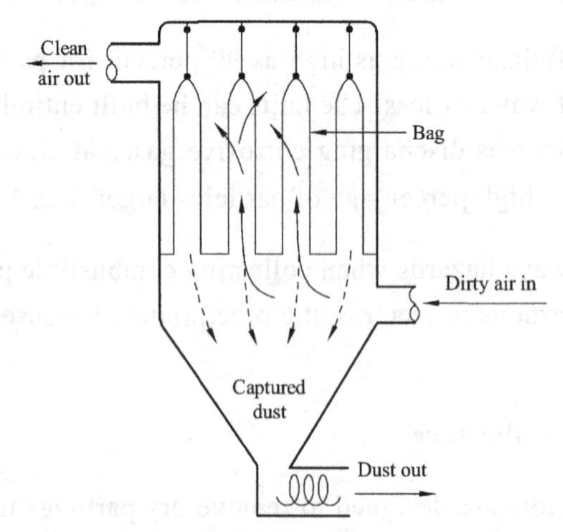

Figure 4.7 A schematic baghouse for control of particulate air pollutants

(Source: P. Aarne Vesilind and Susan M. Morgan, *Introduction to Environmental Engineering*, Thomson Brooks Cole, 2004, P343)

4.2.4 Gas Emission Control

At present, there are four useful ways to reduce emission of undesirable gases: absorption, adsorption, combustion, and condensation.

Absorption The solubility of dissimilar gas in a liquid is different. Control devices based on the principle of absorption attempt to transfer the pollutant from a gas phase to a liquid phase. This is a mass transfer process in which the gas dissolves in the liquid. Mass transfer is a diffusion process wherein the pollutant gas moves from points of higher concentration to points of lower concentration.

Structures such as spray chambers and towers or columns are two classes of devices employed to absorb pollutant gases. In scrubbers(Figure 4.4), which are a type of spray chamber, liquid droplets are used to absorb the gas. In towers, a thin film of liquid is used as the absorption medium. Regardless of the type of device, solubility of the liquid must be relatively high. If water is the solute, it generally limits the application to a few inorganic gases such as NH_3, Cl_2, and SO_2. Scrubbers are relatively inefficient absorbers but have the advantage of being able to simultaneously remove particulates. Towers are much more efficient absorbers but they become plugged by particulate matter.

The amount of absorption that can take place for a nonreactive solution is governed by the partial pressure of the pollutant. A typical absorption equilibrium curve describing the relationship between the partial pressure of a gas over a liquid and the concentration of gas in the liquid is shown in Figure 4.8. In the figure, C^* is the equilibrium concentration of gas molecules in the absorbing liquid corresponding to a partial pressure p of a gas above the liquid, and p^* is the equilibrium partial pressure in the bulk gas corresponding to a concentration C of the gas molecules in the liquid. In industrial scrubbers, flowing bulk liquid contacts the gas mixtures in which the partial pressure of the gas to be removed is p. The driving gradient of pressure or concentration moving the gas into the liquid is then $p - p^*$ or $C^* - C$, respectively. By analogy with heat transfer, an expression describing the absorption rate into the liquid for the gas is

$$\frac{\mathrm{d}N}{\mathrm{d}t} = K_{\mathrm{G}}(p - p^*)A = K_{\mathrm{L}}(C^* - C)A \tag{4-17}$$

where K_G and K_L are empirical coefficients whose values depend on the gas-liquid combination, flow patterns and turbulence in the scrubber, temperature, and other factors; N is the number of molecules of the gas being transferred, and A is the area of contact between the liquid and the gas. Henry's law is a special case of this equation where the equilibrium concentration curve is linear.

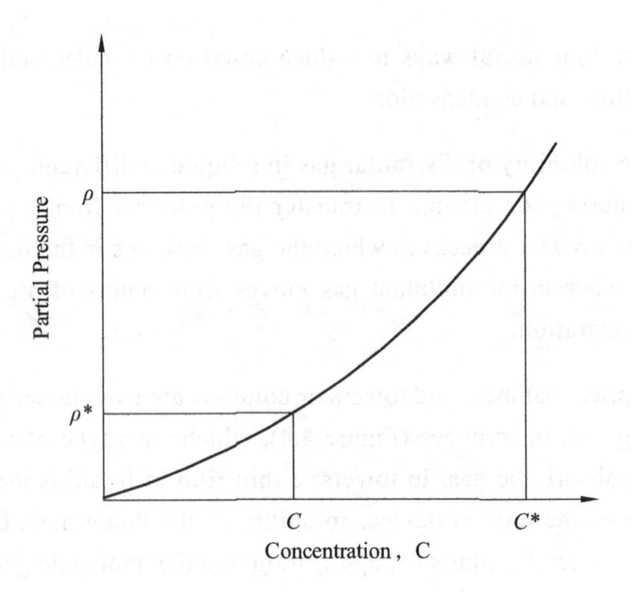

Figure 4.8 A typical absorption equilibrium curve

(Source: J. Glynn Henry and Gary W. Heinke, *Environmental Science And Engineering*, Prentice Hall, 1989, P512)

Adsorption This is a mass-transfer process in which the gas is bonded to a solid. It is a surface phenomenon. The gas (the adsorbate) penetrates into the pores of the solid (the adsorbent) but not into the lattice itself. The bond may be physical or chemical. Electrostatic forces hold the pollutant gas when physical bonding is significant. Chemical bonding is by reaction with the surface. Pressure vessels having a fixed bed are used to hold the adsorbent.

Active carbon (activated charcoal), molecular sieves, silica gel, and activated alumina are the most common adsorbents. Active carbon is manufactured from nut shells (coconuts are great) or coal subjected to heat treatment in a reducing atmosphere. Molecular sieves are dehydrated zeolites (alkali-metal silicates). Sodium silicate is reacted with sulfuric acid to make silica gel. Activated alumina is a porous hydrated aluminum oxide. The common property of these adsorbents is a large "active" surface area per unit volume after treatment. The surface area of the adsorbents varies from 500 to 1,500 m^2 per g for activated carbon to 175 m^2 per g for silica gel. They are very effective for hydrocarbon pollutants. In addition, they can capture H_2S and SO_2. One special form of molecular sieve can also capture NO_2. The amount of adsorbate which the solid can take up is a function of the chemical and physical properties of the solid, particularly the surface area of the pores and fissures in the solid particles within which the gas molecules are deposited. Table 4.3 shows the adsorptive capacity of activated charcoal for three common vapors.

Table 4.3 Adsorptive capacity and retentivity of activated charcoal for three common vapors

Substance	Adsorptive capacity weight (%)	Retention after removal weight (%)
Carbon Tetrachloride	80-110	27-30*
Gasoline	10-20	2-3*
Methanol	50	1.2**

* Regenerated by passing air at 25 °C through the bed for 6 hours.

** Regenerated using steam at 150 °C for 1 hour.

(Source: J. Glynn Henry and Gary W. Heinke, *Environmental Science And Engineering*, Prentice Hall, 1989, P513)

With the exception of the active carbons, adsorbents have the drawback that they preferentially select water before any of the pollutants. Thus, water must be removed from the gas before it is treated. All of the adsorbents are subject to destruction at moderately high temperatures. In industrial processes, the adsorbent is frequently regenerated by passing hot steam through the bed, allowing the water molecules to displace the gas molecules at elevated temperatures (150 °C for active carbon, 600 °C for molecular sieves, 400 °C for silica gel, and 500 °C for activated alumina). The concentrated gas may be recovered, dried, and reprocessed to yield a salable by-product, and the adsorbent is recycled.

Combustion When the contaminant in the gas stream is oxidizable to an inert gas, combustion is a possible alternative method of control. Typically, CO and hydrocarbons fall into this category. Both direct flame combustion by afterburners and catalytic combustion have been used in commercial applications.

Direct flame incineration is the method of choice if two criteria are satisfied. First, the gas stream must have a net heating value(NHV) greater than 3.7 MJ/m^3 . At this NHV, the gas flame will be autogenous (self-supporting after ignition). Below this point, supplementary fuel is required. The second requirement is that none of the by-products of combustion is toxic. In some cases the combustion by-product may be more toxic than the original pollution gas. For example, the combustion of trichloroethylene produces phosgene, which was used as a poison gas in World war Ⅰ. Direct flame incineration has been successfully applied to varnish-cooking, meat-smokehouse, and paint bake-oven emissions.

Some catalytic materials enable oxidation to be carried out in gases that have an NHV of less than 3.7 MJ/m^3 . Conventionally, the catalyst is placed in beds similar to adsorption beds. Frequently, the active catalyst is a platinum or palladium compound. The supporting lattice is usually a ceramic. Aside from expense, a major drawback of the catalysts is their susceptibility to poisoning by sulfur and lead compounds in trace amounts. Catalytic combustion has successfully been applied to printing-press, varnish-cooking, and asphalt-oxidation emission.

4.3 AIR POLLUTION METEOROLOGY

4.3.1 The Atmospheric Engine

Temperature profile Earth's atmosphere can be divided into easily recognizable strata, depending on the temperature profile. Figure 4.9 shows a typical temperature profile for four major layers. The troposphere, where most of our weather occurs, ranges from about 5 km at the poles to about 18 km at the equator. The temperature here decreases with altitude. Over 80% of the air is within this well-mixed layer. On top of the troposphere is the stratosphere, a layer of air where the temperature profile is inverted and in which little mixing takes place. Pollutants that migrate up to the stratosphere can stay there for many years. The stratosphere has a high ozone concentration, and the ozone adsorbs the sun's short-wave ultraviolet radiation. Above the stratosphere are two more layers, the mesosphere and the thermosphere, which contain only about 0.1% of the air.

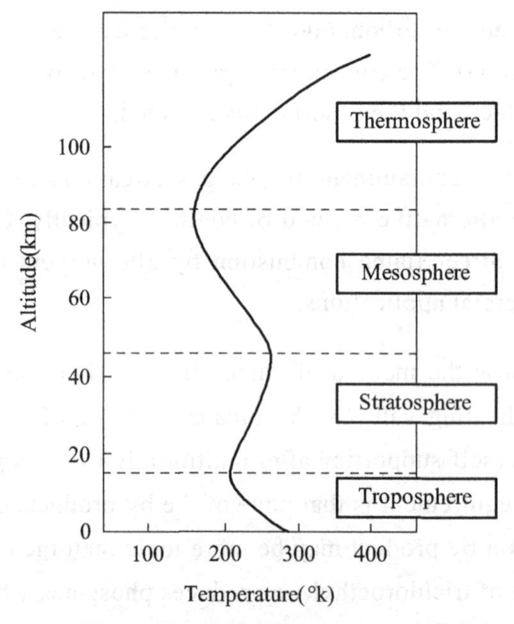

Figure 4.9 Earth's atmosphere

(Source: P. Aarne Vesilind and Susan M. Morgan, *Introduction to Environmental Engineering*, Thomson Brooks Cole, 2004, P303)

The atmosphere is somewhat like an engine. It is continually expanding and compressing gases, exchanging heat, and generally raising chaos. The driving energy for this unwieldy machine comes from the sun. The difference in heat input between the equator and the poles provides the initial overall circulation of the earth's atmosphere. The rotation of the earth coupled with different heat conductivities of oceans and land produce weather.

Wind Because air has mass, it also exerts pressure on things under it. Like water, which we intuitively understand to exert greater pressures at greater depths, the atmosphere exerts more pressure at the surface than it does at higher elevations. The highs and lows depicted on weather maps are simply areas of greater and lesser pressure. The elliptical lines shown on more detailed weather maps are lines of constant pressure (isobars). A two-dimensional plot of pressure and distance through a high or low pressure system would appear as shown in Figures 4.10, a and b, respectively.

The wind flows from the higher pressure areas to the lower pressure areas. On a nonrotating planet, the wind direction would be perpendicular to the isobars [Figure 4.11(a)]. However, since the earth rotates, an angular thrust is added to this motion (the Coriolis effect). The resultant wind direction in the northern hemisphere is as shown in Figure 4.11(b). The technical names given to these systems are anticyclones for highs and cyclones for lows. Anticyclones are associated with good weather. Cyclones are associated with foul weather. Tornadoes and hurricanes are the foulest of the cyclones.

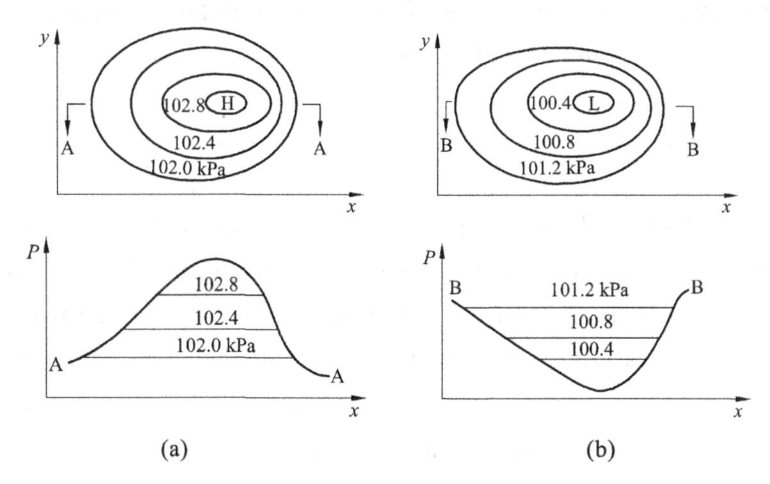

Figure 4.10 High and low pressure systems

(Source: Mackenzie L. Davis and David A. Cornwell, *Introduction to Environmental Engineering*, Tsinghua University Press, 2007, P581)

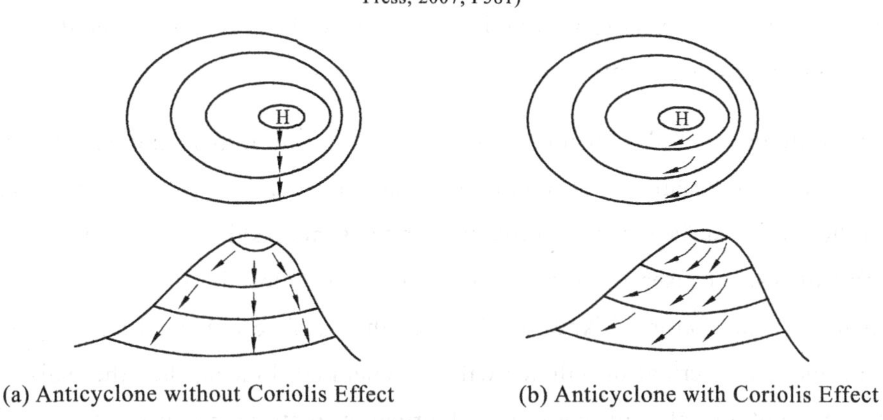

(a) Anticyclone without Coriolis Effect (b) Anticyclone with Coriolis Effect

Figure 4.11 Wind flow due to pressure gradient

(Source: Mackenzie L. Davis and David A. Cornwell, *Introduction to Environmental Engineering*, Tsinghua University Press, 2007, P581)

Wind speed is in part a function or the steepness of the pressure surface. When the isobars are close together, the pressure gradient (slope) is said to be steep and the wind speed relatively high. If the isobars are well spread out, the winds are light or nonexistant.

Observations indicate that wind speed varies with height above the ground as a function of atmospheric stability and surface roughness. The variation of wind speed with height is frequently described using the power law

$$u_z = u_{10}\left(\frac{Z}{Z_{10}}\right)^P \tag{4-18}$$

where u_z = wind speed at height Z above the ground

u_{10} = wind speed at 10 m (measurement height specified by World Meteorological Organization for meteorological stations)

P = exponent depending upon atmospheric stability and the character of the underlying surface (varies from about 0.1 to 0.4).

4.3.2 Turbulence

Mechanical turbulence In its simplest terms, we may consider turbulence to be the addition of random fluctuations of wind velocity (that is, speed and direction) to the overall average wind velocity. These fluctuations are caused, in port, by the fact that the atmosphere is being sheared. The shearing results from the fact that the wind speed is zero at the ground surface and rises with elevation to near the speed imposed by the pressure gradient. The shearing results in a tumbling, tearing motion as the mass just above the surface falls over the slower moving air at the surface. The swirls thus formed are called eddies. These small eddies feed larger ones. As you might expect, the greater the mean wind speed, the greater the mechanical turbulence. The more mechanical turbulence there is, the easier it is to disperse and spread atmospheric pollutants.

Thermal turbulence Like all other things in nature, the rather complex interaction that produces mechanical turbulence is confounded and further complicated by a third party. Heating of the ground surface causes turbulence in the same fashion that heating the bottom of a beaker full of water causes turbulence. At some point below boiling, you can see density currents rising off the bottom. Likewise, if the earth's surface is heated strongly and in turn heats the air above it, thermal turbulence will be generated. Indeed, the "thermals" sought by glider pilots and hot air balloonists are these thermal currents rising on what otherwise would be a calm day.

The converse situation can arise during clear nights when the ground radiates its heat away to the cold night sky. The cold ground, in turn, cools the air above it, causing a sinking density current.

4.3.3 Stability

The tendency of the atmosphere to resist or enhance vertical motion is termed stability. It is related to both wind speed and the change of air temperature with height (lapse rate). For our purpose, we may use the lapse rate alone as an indicator of the stability condition of the atmosphere.

There are three stability categories. When the atmosphere is classified as unstable, mechanical turbulence is enhanced by the thermal structure. A neutral atmosphere is one in which the thermal structure neither enhances nor resists mechanical turbulence. When the thermal structure inhibits mechanical turbulence, the atmosphere is said to be stable.

Cyclones are associated with unstable air. Anticyclones are associated with stable air.

Neutral stability The lapse rate for a neutral atmosphere is defined by the rate of temperature increase (or decrease) experienced by a parcel of air that expands (or contract) adiabatically (without the addition or loss of heat) as it is raised through the atmosphere. This rate of temperature decrease ($\mathrm{d}T/\mathrm{d}Z$) is called the ***dry adiabatic lapse rate***. It is designated by the Greek letter gamma (Γ). It has a value of approximately -1.00 °C/100 m. In Figure 4.12(a), the dry adiabatic lapse rate of a parcel of air is shown as a dashed line and the temperature of the atmosphere (ambiet lapse rate) is shown as a solid line. Since the ambient lapse rate is the same as Γ, the atmosphere is said to have a neutral stability.

Unstable atmosphere If the temperature of the atmosphere falls at a rate greater than Γ, the lapse rate is said to be ***superadiabatic***, and the atmosphere is unstable. Using Figure 4.12(b), we can see that this is so. The actual lapse rate is shown by the solid line. If we capture a balloon full of polluted air at elevation A and adiabatically displace it 100 m vertically to elevation B, the temperature of the air inside the balloon will decrease from 21.15 to 20.15 °C. At a lapse rate of -1.25 °C/100 m, the temperature of the air outside the balloon will decrease from 21.15 to 19.90 °C. The air inside the balloon will be warmer than the air outside. This temperature difference gives the balloon buoyancy. It will behave as a hot gas and continue to rise without any further mechanical effort. Thus, mechanical turbulence is enhanced and the atmosphere is unstable. If we adiabatically displaced the balloon downward to elevation C, the temperature inside the balloon would rise at the rate of the dry adiabat. Thus, in moving 100 m, the temperature will increase from 21.15 to 22.15 °C. The temperature outside the balloon will increase at the superadiabatic lapse rate to 22.40 °C. The air in the balloon will be cooler than the ambient air and the balloon will have a tendency to sink. Again, mechanical turbulence (displacement) is enhanced.

Stable atmosphere If the temperature of the atmosphere falls at a rate less than Γ, it is called ***subadiabatic***. The atmosphere is stable. If we again capture a balloon of polluted air at elevation A [Figure 4.12(c)] and adiabatically displace it vertically to elevation B, the temperature of the polluted air will decrease at a rate equal to the dry adiabatic rate. Thus, in moving 100 m, the temperature will decrease from 21.15 to 20.15 °C as before. However, since the ambient lapse rate is – 0.5 °C/100 m, the temperature of the air outside the balloon will drop to only 20.65 °C. Since the air inside the balloon is cooler than the air outside the balloon, the balloon will have a tendency to sink. Thus, the mechanical displacement (turbulence) is inhibited.

In contrast, if we displaced the balloon adiabatically to elevation C, the temperature inside the balloon would increase to 22.15 °C, while the ambient temperature would increase to 21.65 °C. In this case, the air inside the balloon would be warmer than the ambient air and the balloon would tend to rise. Again, the mechanical displacement would be inhibited.

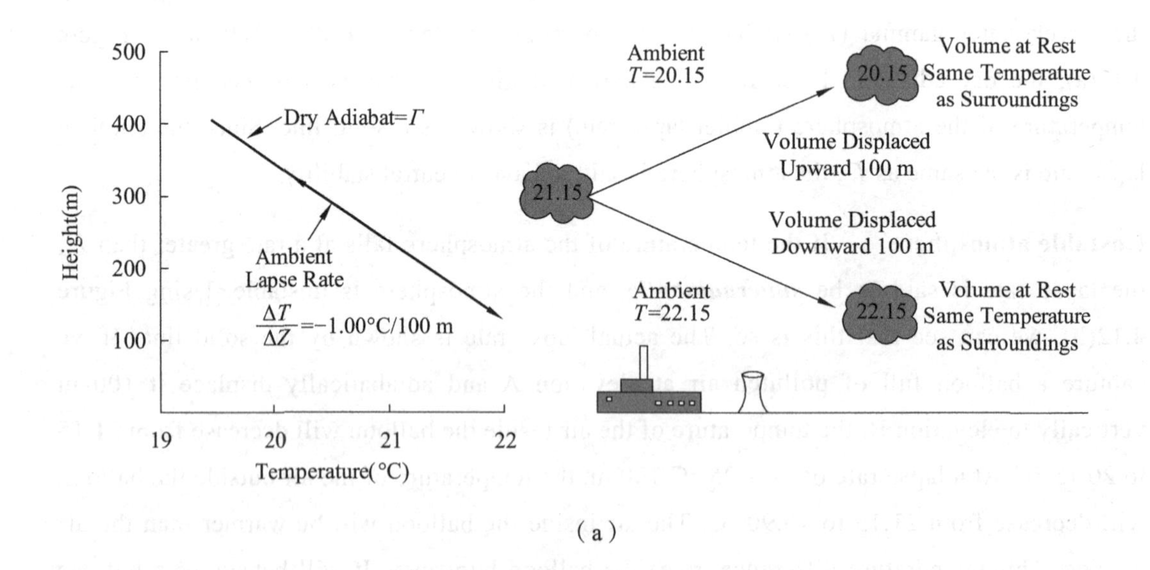

(a)

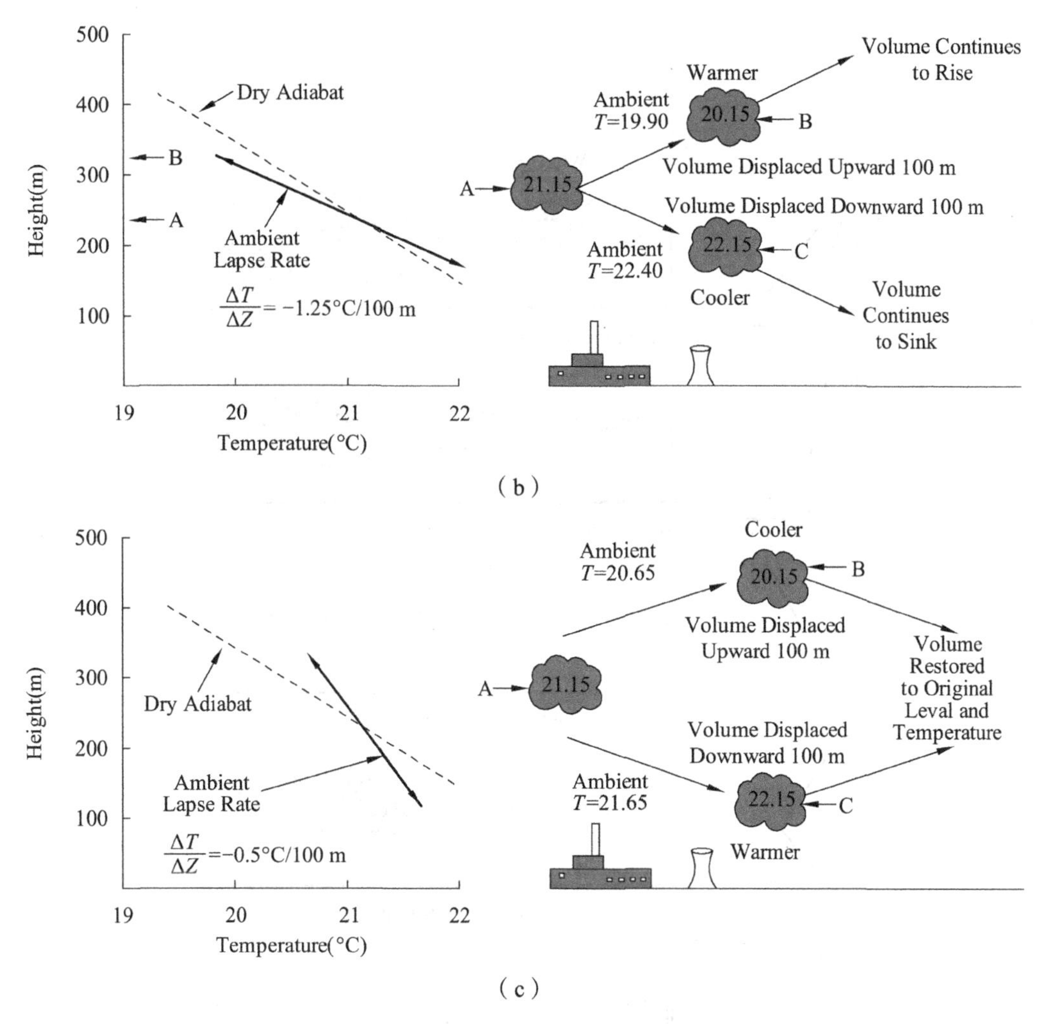

Figure 4.12 Lapse rate and displaced air volume

(Source: Mackenzie L. Davis and David A. Cornwell, *Introduction to Environmental Engineering*, Tsinghua University Press, 2007, P584)

There are two special cases of subadiabatic lapse rate. When there is no change of temperature with elevation, the lapse rate is called ***isothermal***. When the temperature increases with elevation, the lapse rate is called an ***inversion***. The inversion is the most severe form of a stable temperature profile. It is often associated with restricted air volumes, which cause air pollution episodes.

Plume types The smoke trail or plume from a tall stack located on flat terrain has been found to exhibit a characteristic shape that is dependent on the stability of the atmosphere. The six classical plumes are shown in Figure 4.13, along with the corresponding temperature profiles. In each case, Γ is given as a broken line to allow comparison with the actual lapse rate, which is given as a solid line. In the bottom three cases, particular attention should be given to the location of the inflection point with respect to the top of the stack.

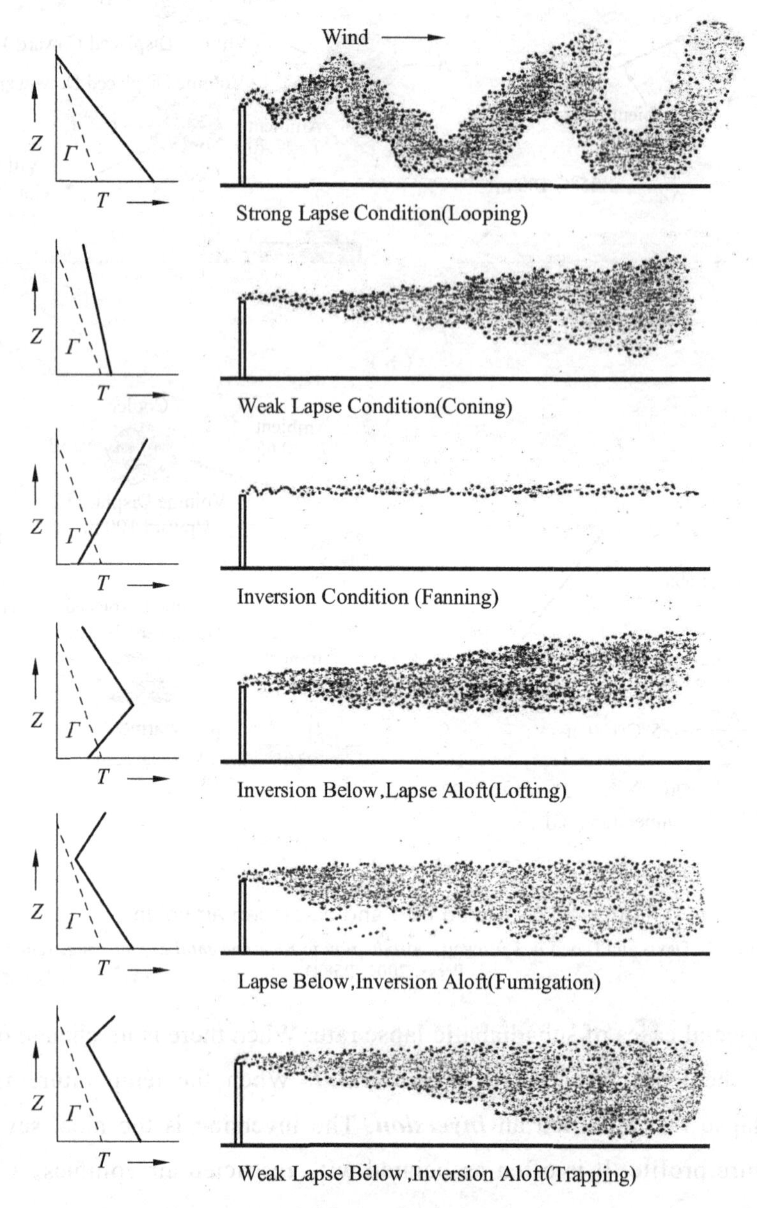

Figure 4.13 Six types of plume behavior

(Source: Mackenzie L. Davis and David A. Cornwell, *Introduction to Environmental Engineering*, Tsinghua University Press, 2007, P586)

4.3.4 Terrain Effects

Heat islands A heat island results from a mass of material, either natural or anthropogenic, that absorbs and reradiates heat at a greater rate than the surrounding area.

This causes moderate to strong vertical convection currents above the heat island. The effect is superimposed on the prevailing meteorological conditions. It is nullified by strong winds. Large industrial complexes and small to large cities are examples of places that would have a heat island.

Because of the heat island effect, atmospheric stability will be less over a city than it is over the surrounding countryside. Depending upon the location of the pollutant sources, this can be either good news or bad news. First, the good news: For ground level sources such as automobiles, the bowl of unstable air that forms will allow a greater air volume for dilution of the pollutants. Now the bad news: Under stable conditions, plumes from tall stacks would be carried out over the countryside without increasing ground level pollutant concentrations. Unfortunately, the instability caused by the heat island mixes these plumes to the ground level.

Land/sea breezes Under a stagnating anticyclone, a strong local circulation pattern may develop across the shoreline of large water bodies. During the night, the land cools more rapidly than the water. The relatively cooler air over the land flows toward the water(a land breeze, Figure 4.14). During the morning the land heats faster than water. The air over the land becomes relatively warm and begins to rise. The rising air is replaced by air from over the water body (a sea or lake breeze, Figure 4.15).

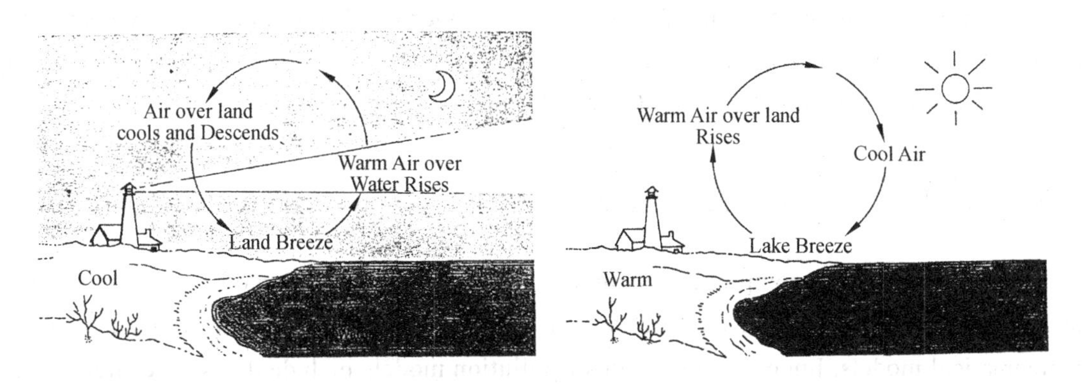

Figure 4.14 Land breeze during the night Figure 4.15 Lake breeze during the day

(Source: Mackenzie L. Davis and David A. Cornwell, *Introduction to Environmental Engineering*, Tsinghua University Press, 2007, P587)

The effect of the lake breeze on stability is to impose a surface-based inversion on the temperature profile. As the air moves from the water over the warm ground, it is heated from below. Thus, for stack plumes originating near the shore line, the stable lapse rate causes a fanning plume close to the stack (Figure 4.16). The lapse condition grows to the height of the stack as the air moves inland. At some point inland, a fumigation plume results.

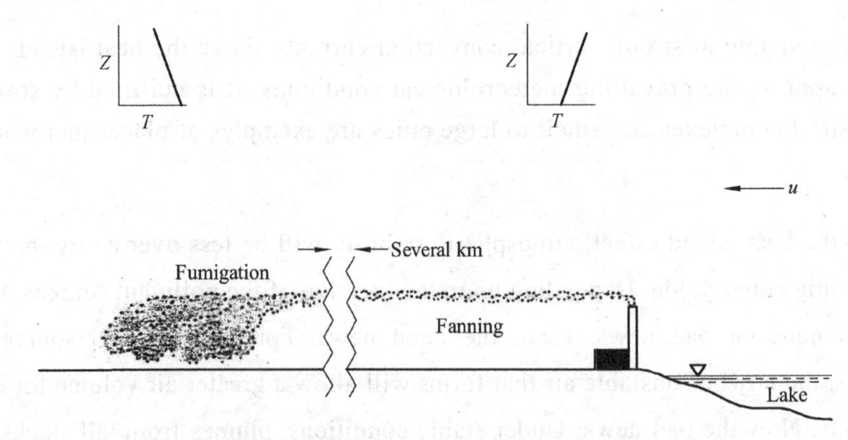

Figure 4.16 Effect of lake breeze on plume dispersion

(Source: Mackenzie L. Davis and David A. Cornwell, *Introduction to Environmental Engineering*, Tsinghua University Press, 2007, P588)

Valleys When the general circulation imposes moderate to strong winds, valleys that are oriented at an acute angle to the wind direction channel the wind. The valley effectively peels off part of the wind and forces it to follow the direction of the valley floor.

Under a stagnating anticyclone, the valley will set up its own circulation. Warming of the valley walls will cause the valley air to be warmed. It will become more buoyant and flow up the valley. At night the cooling process will cause the wind to flow down the valley.

Valleys are more susceptible to inversions than level terrain. The valley walls protect the floor from radiative heating by the sun. Yet the walls and floor are free to radiate away heat to the cold night sky. Thus, under weak winds, the ground cannot heat the air rapidly enough during the day to dissipate the inversion that formed during the night.

4.4 PREDICTING AIR POLLUTANT CONCENTRATIONS

The design of industrial complexes, community planning, identification of significant sources, and prediction of pollutant concentrations at selected receptors are usually done by using mathematical models. Important inputs to air pollution models include the type, character, and distribution of the sources; the pollutants emitted; meteorological variables which determine the transport and dispersion of pollutants; and the chemical reactions of pollutants in the atmosphere. In the near field (less than 20 km from the source), except for selected pollutants such as fluorine, H_2S, and photochemical oxidants, we can usually neglect atmospheric chemical reactions and removal processes. In the far field (greater than 100 km from the source), chemical reactions and removal processes become increasingly important. Because models are greatly simplified representations of actual processes, single-valued predictions should be regarded as being within a factor of two at best.

4.4.1 Pollution Dispersion Models

Classification of Dispersion models A dispersion model is a mathematical description of the meteorological transport and dispersion process that is quantified in terms of source and meteorologic parameters during a particular time. The resultant numerical calculations yield estimates of concentrations of the particular pollutant for specific locations and times.

The dispersion models are generally classified as either short-term or climatological models. Short-term models are generally used under the following circumstances: 1) to estimate ambient concentrations where it is impractical to sample, such as over rivers or lakes, or at great distances above the ground; 2) to estimate the required emergency source reductions associated with period of air stagnations under air pollution episode alert conditions; and 3) to estimate the most probable locations of high, short-term, ground-level concentrations as part of a site selection evaluation for the location of air monitoring equipment.

Climatological models are used to estimate mean concentrations over a long period of time or to estimate mean concentrations that exist at particular times of the day for each season over a long period of time. Long-term models are used as an aid for developing emissions standards. We will be concerned only with short-term models in their most simple application.

Gaussian dispersion model By making many simplifying assumptions, Gaussian diffusion equations can be developed to describe the atmospheric dispersion of a puff in three dimensions or a steady-state plume from a continuous source in two dimensions. For the simplest model, we assume that a plume traveling horizontally (in the x direction) at a mean speed $\overline{u}$ disperses horizontally (y) and vertically (z) so that the concentration of a pollutant at any cross section of the plume follows the normal (Gaussian) probability distribution. If also, for any point (x, y, z) in the plume, the concentration C of pollutant at that point is such that

$$C_{(x,y,z)} \alpha \frac{1}{\overline{u}} \quad (\overline{u} = \text{average wind speed})$$

$$C_{(x,y,z)} \alpha Q \quad (Q = \text{source strength})$$

and

$$C_{(x,y,z)} \alpha G \quad (G = \text{normalized Gaussian curve in the } y \text{ and } z \text{ directions})$$

then

$$C_{(x,y,z)} = \frac{Q}{\overline{u}} G_y G_z$$

The expression for the Gaussian function G_y normalized so that the area under the curve is unity is

$$G_y = \frac{1}{\sqrt{2\pi}\sigma_y} \exp\left[-\frac{1}{2}\left(\frac{y}{\sigma_y}\right)^2\right]$$

and similarly for G_z so that

$$C_{(x,y,z)} = \frac{Q}{\bar{u}}\left\{\frac{1}{\sqrt{2\pi}\sigma_y} \exp\left[-\frac{1}{2}\left(\frac{y}{\sigma_y}\right)^2\right]\right\}\left\{\frac{1}{\sqrt{2\pi}\sigma_z} \exp\left[-\frac{1}{2}\left(\frac{z}{\sigma_z}\right)^2\right]\right\}$$

$$= \frac{Q}{2\pi\sigma_y\sigma_z\bar{u}} \exp\left[-\frac{1}{2}\left(\frac{y}{\sigma_y}\right)^2\right] \exp\left[-\frac{1}{2}\left(\frac{z}{\sigma_z}\right)^2\right] \quad (4\text{-}19)$$

where σ_y and σ_z are the standard deviations of the dispersion in the y and z directions, respectively, x=0 at the source and y and z are zero on the plume center line.

In order to relate this expression to the ground level rather than to the centerline of the plume, we can make the height of any point C in the plume a distance Z above the ground. In this case, the vertical height of point C above the centerline of the plume becomes $Z - H$ (see Figure 4.17) and the equation becomes

Cr and Ci=concentration due to real and imaginary sources, respectively.

$$C_{(x,y,z)} = \frac{Q}{2\pi\sigma_y\sigma_z\bar{u}} \exp\left[-\frac{1}{2}\left(\frac{y}{\sigma_y}\right)^2\right] \exp\left[-\frac{1}{2}\left(\frac{Z-H}{\sigma_z}\right)^2\right] \quad (4\text{-}20)$$

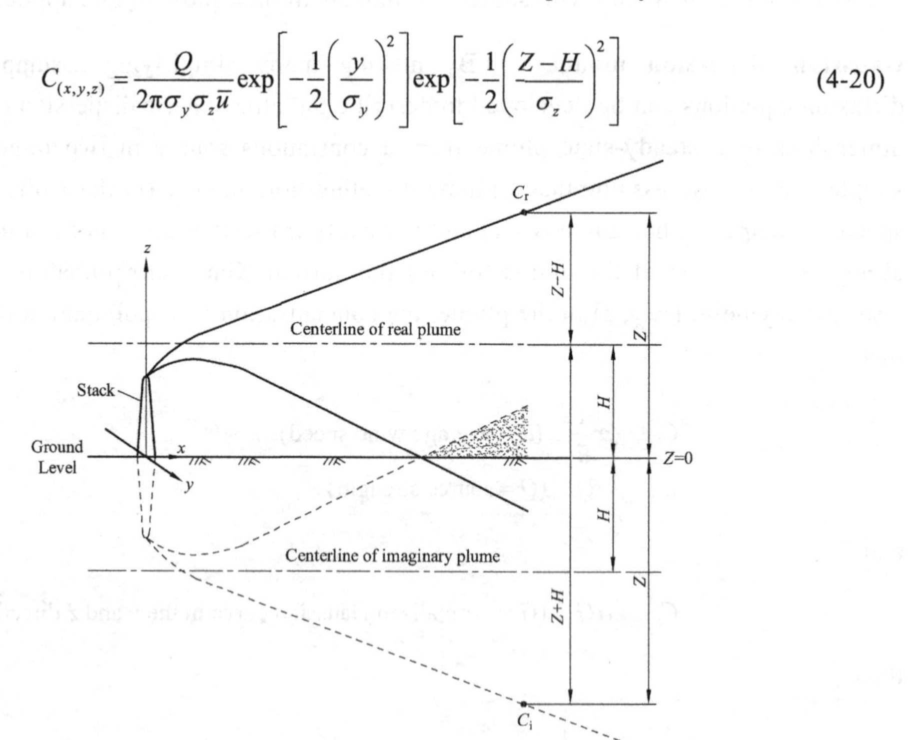

Figure 4.17 Sketch of plume dispersion in the vertical direction and the reflection of pollutants at ground level

(Source: J. Glynn Henry and Gary W. Heinke, *Environmental Science And Engineering*, Prentice Hall, 1989, P527)

Provided that the plume does not impinge on the ground, this model should apply. However, because the ground tends to reflect rather than remove pollutants, a technique assuming 100 percent reflection of pollutants is used to account for the increased pollutant concentration at ground level (see Figure 4.17 again). A mirror image of the plume is envisaged, and the concentration of pollutant at an imaginary point (at a location $Z + H$) is added to the concentration in the real plume. The plume diffusion equation, in its most common form, then becomes

$$C=\frac{Q}{2\pi\sigma_y\sigma_z\bar{u}}\exp\left[-\frac{1}{2}\left(\frac{y}{\sigma_y}\right)^2\right]\left\{\exp\left[-\frac{1}{2}\left(\frac{Z-H}{\sigma_z}\right)^2\right]+\exp\left[-\frac{1}{2}\left(\frac{Z+H}{\sigma_z}\right)^2\right]\right\} \tag{4-21}$$

where

C = the pollutant concentration (kg/m^3) at a receptor located at (x, y, z)

σ_y and σ_z = the standard deviations of the dispersion in the y and z directions, respectively (m)

$\bar{u}$ = the mean wind speed through the layer in which diffusion takes place (m/s)

x, y, and z = spatial coordinates of the receptor (m) relative to the source (the x axis is oriented in the direction of the mean wind, y is at right angles to x in the horizontal plane, z is in the vertical plane, and Z is the vertical coordinate relative to ground level)

H = the effective height of the pollutant release (m)

Q = the source emission rate (kg/s)

Some of the assumptions made to develop this Guassian diffusion equation are as follows:

1. All of the pollutants are emitted from a point source of infinite strength.

2. The concentration distribution across the width and depth of the plume is Gaussian.

3. The wind is uniform through the layer in which dispersion occurs, and an average wind can be used in the equation. In practice, the wind used in the equation is taken to be the wind at the top of a stack for an elevated source, estimated using Equation 4-18.

4. The pollutant under consideration is not lost by decay, chemical reaction, or deposition; It is supposed that all of the pollutant which impinges at the earth's surface is fully "reflected".

5. The equation is to be used over relatively flat, homogeneous terrain.

6. The equation represents a steady state solution ($2Q/2t = 0$)over the averaging period.

7. The pollutants have the same density as the air surrounding them.

Note how the equation reduces to a simpler form for concentrations at specific ground-level locations, such as at a distance y from the centerline ($Z = 0$) or on the centerline of the plume (y=0, Z=0).

The values of σ_y and σ_z depend upon the turbulent structure or stability of the atmosphere. Figure 4.18 and Figure 4.19 provide graphical relationships between the downwind distance x in kilometers and values of σ_y and σ_z in meters. The curves on the two figures stand for the stability categories labeled "A" through "F". Table 4.4 describes the method for determining the stability categories based on wind speed, time of day (radiation), and cloud cover. Category A corresponds to an extremely unstable atmosphere conditions, B to moderately unstable atmosphere conditions, C to slightly unstable to neutral conditions, D to neutral conditions, E to slightly stable atmosphere conditions, and F to a moderately stable atmospheric conditions. The curves in Figure 4.18 are for continuous point-source plumes over averaging periods of 10 minutes or so. They should not be used to describe the diffusion of a puff in three dimensions.

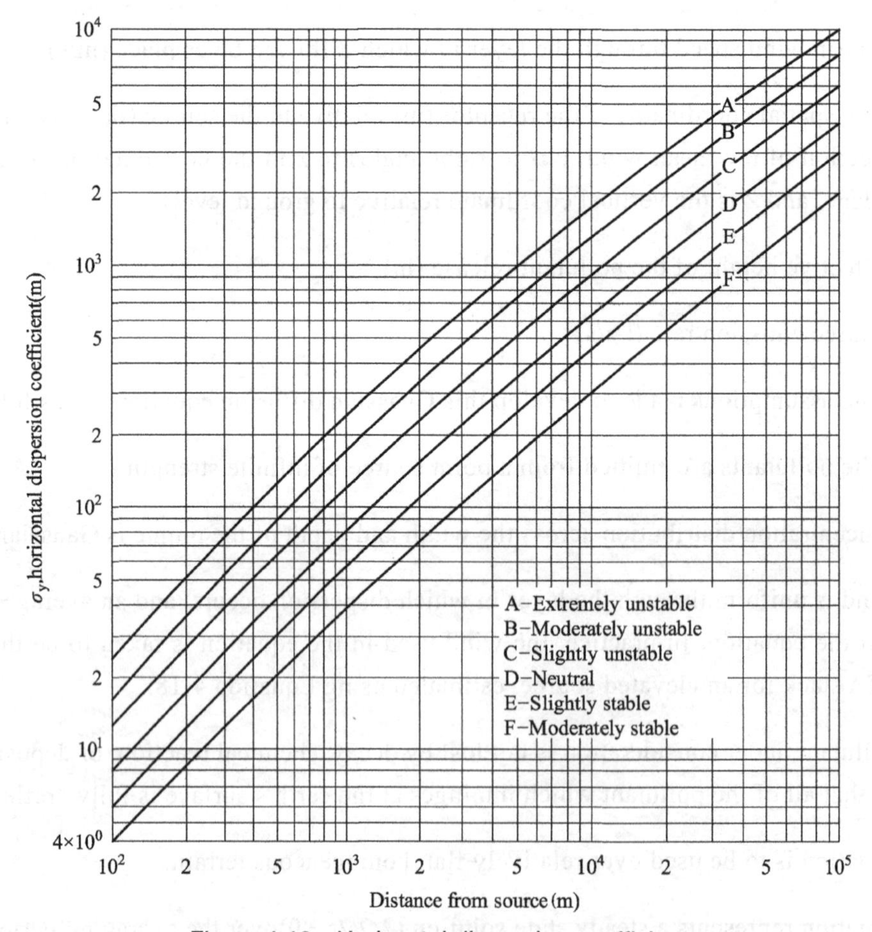

Figure 4.18 Horizontal dispersion coefficient

(Source: Mackenzie L. Davis and David A. Cornwell, *Introduction to Environmental Engineering*, PWS Engineering, 1985, P359)

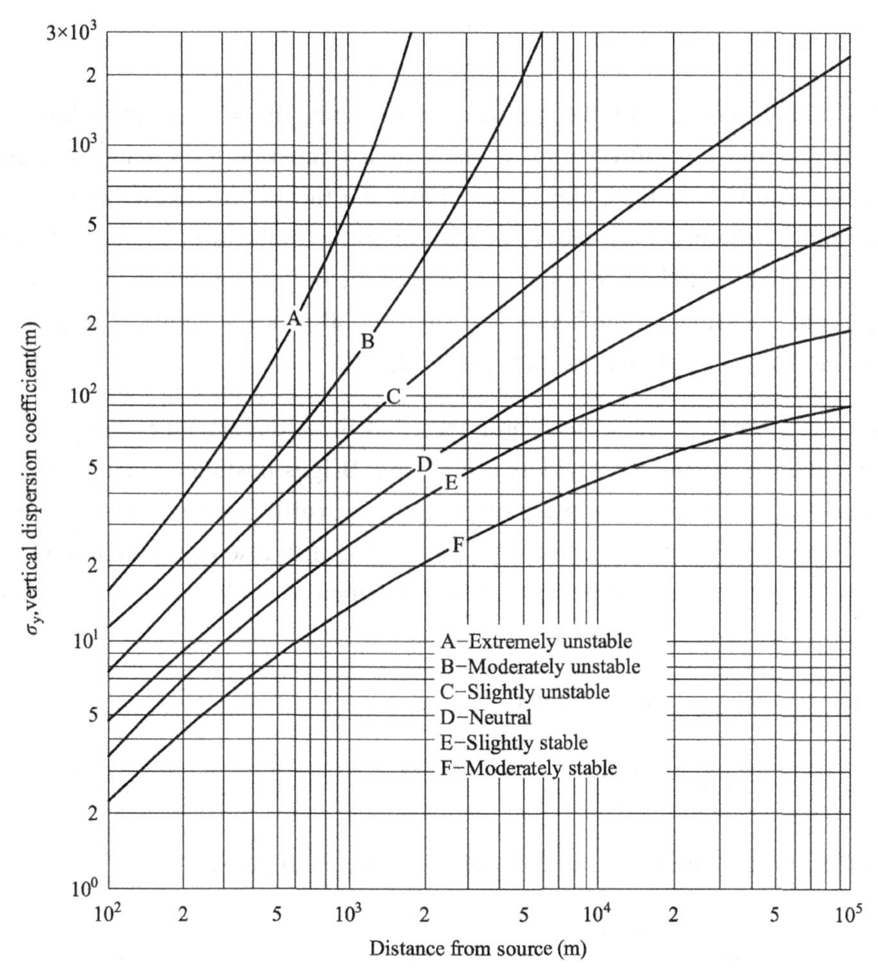

Figure 4.19 Vertical dispersion coefficient

(Source: Mackenzie L. Davis and David A. Cornwell, *Introduction to Environmental Engineering*, PWS Engineering, 1985, P360)

Table 4.4 Key to stability categories

Surface wind speed (at 10 m) (m/s)	Day			Night	
	Incoming solar radiation			Amount of overcast	
	Strong	Moderate	Slight	⩾4/8 Low cloud	⩽3/8 Low cloud
<2	A	A-B	B		
2-3	A-B	B	C	E	F
3-5	B	B-C	C	D	E
5-6	C	C-D	D	D	D
>6	C	D	D	D	D

Note: Categories A, B, C range from extremely unstable to slightly unstable. Category D is neutral and should be assumed for overcast conditions during day or night. Categories E, F indicate slightly stable and moderately stable conditions respectively. Adapted from Turner, 1970.

(Source: Mackenzie L. Davis and David A. Cornwell, *Introduction to Environmental Engineering*, Tsinghua University Press, 2007, P594)

4.4.2 Plume Rise Models

The final variable required in the application of Equation 4-21 is the plume height H. Observation of a plume emitted from a stack at a temperature T_s above the ambient air temperature T_A shows that the plume rises above the top of the stack due to its discharge momentum and its thermal buoyancy. For plumes from combustion sources, the momentum rise is so small relative to the buoyancy rise (due to the high temperature of the plume) that it can be neglected. The final plume height H is the sum of the stack height H_s and the combined momentum and buoyancy plume rise ΔH, i.e. $H=H_s+\Delta H$.

Many plume rise equations have been proposed, but those developed by Briggs(1969) using dimensional analysis are the most widely used today. Briggs postulated that plume rise occurs simultaneously with a relatively rapid plume expansion (diffusion) as a result of entrainment of ambient air into the plume. Therefore, the plume rise must also be a function of the stability of the atmosphere. Briggs proposed the following equations to describe the buoyant rise of a warm plume:

1. For stable and near-neutral conditions,

$$\Delta H = 2.6\left(\frac{F}{\bar{u}S}\right)^{1/3} \tag{4-22}$$

where F is the initial buoyancy flux of the emitted plume defied by

$$F = \frac{g(T_s - T_A)W}{T_s}\left(\frac{D}{2}\right)^2 \tag{4-23}$$

where, in turn, g = acceleration due to gravity (m/s^2)

T_s and T_A = temperatures of the emitted gas and the environment, respectively, at the point of emission (K)

W = exit velocity of the plume (m/s)

D = diameter of the stack at the top (m)

$\bar{u}$ = mean wind speed through the atmospheric layer of depth H, frequently taken to correspond to u at the height H_s of the stack

S = a stability parameter defined by

$$S = \frac{g}{T_A}\left[\frac{\Delta T_A}{\Delta Z} + 0.01\ ^\circ\text{C/m}\right] \tag{4-24}$$

The coefficient 2.6 in Equation 4-22 was determined empirically, and the value of $\Delta T_A/\Delta Z$

through the layer of plume rise should be used to determine S.

2. For unstable atmospheres where the plume theoretically would never stop rising as a result of ambient air entrainment,

$$\Delta H = 1.6 \frac{F^{1/3} x^{2/3}}{\bar{u}} \tag{4-25}$$

For unstable atmospheres, there is no general agreement on where the plume rise should be terminated, but it is reasonable to consider the rise terminated by the time the plume has traveled 10 stack heights or so downstream. (i.e. u=10H_s)

Example 4.3

A proposed paper processing mill is expected to emit 1/2 tonne of H_2S per day from a single stack. The nearest receptor is a small town 1,700 m northeast of the mill site, and southwest winds are expected to occur 15% of the time. The stack at the mill must be sufficiently high that the H_2S concentration in the town will not exceed 20 ppb by volume (28 $\mu g/m^3$ on a mass basis) at the ground. The physical characteristics of the emissions and the ambient atmosphere are as follows:

Gas exit velocity W	= 20 m/s
Gas exit temperature T_s	= 122 °C
Stack diameter D at the top	= 2.5 m
Ambient air temperature T_A	= 17 °C
Wind velocity u assumed for conservative analysis	= 2 m/s
Temperature lapse rate γ	= 6 °C/km (assumed)

Estimate the required stack height at the mill.

Solution Using Equation 4-21, the maximum ground-level (Z=0) concentration will occur on the horizontal centerline (y=0). At y=0,

$$\exp\left(-\frac{y^2}{2\sigma_y^2}\right) = 1$$

and at Z=0,

$$\left\{\exp\left[-\frac{(Z-H)^2}{2\sigma_z^2}\right] + \exp\left[-\frac{(Z+H)^2}{2\sigma_z^2}\right]\right\} = 2\exp\left(-\frac{H^2}{2\sigma_z^2}\right)$$

Thus, Equation 4-21 reduces to the simple form

$$C_{0,0,x} = \frac{Q}{\pi\sigma_y\sigma_z\bar{u}}\left[\exp\left(-\frac{H^2}{2\sigma_z^2}\right)\right]$$

The source strength is

$$Q = \frac{500 \text{ kg/d}}{86\,400 \text{ s/d}} = 5.79\times10^{-3} \text{ kg/d}$$

A temperature lapse rate of 6 °C/km represents a slightly stable atmosphere, say, stability category E.

At x=1,700 m, for stability E, from Figure 4.18 and 4.19, we have σ_y=80 m and σ_z=30 m. Solving for H, we obtain

$$28\times10^{-9} = \frac{5.79\times10^{-3}}{\pi\times80\times30\times2}\left[\exp-\left(\frac{H^2}{2\times(30)^2}\right)\right]$$

$$\exp\left(-\frac{H^2}{1\,800}\right) = \frac{28\times10^{-9}\times3.14\times80\times30\times2}{5.79\times10^{-3}} = 0.729$$

$$H^2 = 1\,800 \ln(0.79) = 569 \text{ m}^2$$

$$H = 23.9 \text{ m}$$

The plume rise ΔH is estimated using Equation 4-22 for a slightly stable atmosphere. For the conditions specified,

$$F = \frac{g(T_s - T_A)W}{T_s}\left(\frac{D}{2}\right)^2 = \frac{9.81\times(395-290)\times20}{395}\left(\frac{2.5}{2}\right)^2 = 81.4 \text{ m}^3/\text{s}^3$$

and from Equation 4-24

$$S = \frac{9.81}{290}[-0.006+0.01] = 1.35\times10^{-4}$$

Therefore from Equation 4-22

$$\Delta H = 2.6\left(\frac{81.4}{2\times1.35\times10^{-4}}\right)^{1/3} = 17.4 \text{ m}$$

Thus, the stack height H_s must be $H - \Delta H = 23.9 - 17.4 = 6.5$ m high.

Note that the prediction of stack height in Example 4.3 is a minimum height for flat terrain and is not sufficiently precise to warrant three-figure accuracy. The design stack height would thus be 7 m for the conditions specified. The actual stack height would be selected after repeating the calculation many times for various meteorological conditions and after consideration of other factors, such as damage to vegetation at various distances downstream.

Chapter 5
Solid Wastes

Solid wastes in the broadest sense include all the discarded solid materials from municipal, industrial, and agricultural activities. However, for the discussion to follow, solid wastes will refer only to those solid and/or semisolid wastes which are the responsibility of, and usually collected by, a municipality. Residential and commercial areas, together with some industrial operations, are the sources of these "nonhazardous" municipal wastes.

In this chapter, some basic concepts and the characteristics of solid wastes are introduced. Landfilling, incineration, composting, and other methods for the disposal of solid wastes are discussed, followed by an introduction of the methods of solid wastes resource.

5.1 CHARACTERIZATION OF SOLID WASTES

5.1.1 Concept of Solid Waste

In general terms, solid waste (sometimes called refuse) can be defined as waste not transported by water, which has been rejected for further use. For municipal solid wastes, more specific terms are applied to the putrescible (biodegradable) food wastes, called ***garbage***, and the nonputrescible solid wastes referred to as ***rubbish***. Trash is synonymous with rubbish in some countries, but trash is technically a subcomponent of rubbish. Rubbish can include a variety of materials that may be combustible (paper, plastic, textiles, etc.) or noncombustible (glass, metal, masonry, etc.). Most of these kinds of waste are discarded on a regular basis from specific locations. However, there are wastes – sometimes called "special wastes" – such as construction debris, leaves and street litter, abandoned automobiles, and old appliances, which are collected at sporadic intervals from different places.

Not included in the components of municipal waste as just described are many other solid wastes that are not normally a municipal responsibility. Such things as ashes from coal-fired generating stations, sludges from water and wastewater treatment plants, wastes from animal feedlots, mine tailings, and other industrial solid wastes are in this category and require separate arrangements for their disposal.

5.1.2 Quantities of Solid Waste

There are wide differences in amounts collected by municipalities because of differences in climate, living standards, time of year, education, location, and collection and disposal practices. In 2003, the EPA of United States estimated that the national average rate of solid waste generated was 2.04 kg/capita day. On this basis, in 2003, the U.S. produced 214 teragrams (Tg) of solid waste ($1\text{Tg} = 1 \times 10^9$ kg). This is a 56 percent increase over the 1980 estimate of 137.8 Tg and a nearly 170 percent increase over the 1960 estimate of 80.1 Tg. The EPA estimates that 60 percent of the waste stream comes from residential sources, and the remainder is from commercial sources. Individual cities may vary greatly from these estimates. For example, Los Angeles, California, generates about 3.18 kg/capita day while the rural community of Wilson, Wisconsin, generates about 1.0 kg/capita day.

In wet areas, because of moisture absorbed by solid waste, the amount collected may exceed the amount generated, which is usually reported on a dry basis. On the other hand, with the use of home grinders, on-site storage of recyclable materials, and other conservation measures, the amount collected may be less than that generated. In this chapter, differences between the quantities generated and collected are ignored because variations due to other factors are much more significant.

5.1.3 Characteristics of Solid Waste

Classifications The common materials of solid waste can be classified in several different ways. The point of origin is important in some cases, so classification as domestic, institutional, commercial, industrial, street, demolition, or construction may be useful. The nature of the material may be important, so classification can be made on the basis of organic, inorganic, combustible, noncombustible, putrescible, and nonputrescible fractions. One of the most useful classifications is based on the kinds of materials as shown in table 5.1.

Sources Garbage is the animal and vegetable waste resulting from the handling, preparation, cooking, and serving of food. It is composed largely of putrescible organic matter and moisture; it includes a minimum of free liquids. The term does not include food processing wastes from canneries, slaughterhouses, packing plants, and similar facilities, or large quantities of condemned food products. Garbage originates primarily in home kitchens, stores, markets, restaurants, and other places where food is stored, prepared, or served. Garbage decomposes rapidly, particularly in warm weather, and may quickly produce disagreeable odors. There is some commercial value in garbage as animal food and as a base for commercial feeds. However, this use may be precluded by health considerations.

Rubbish consists of a variety of both combustible and noncombustible solid wastes from homes, stores, and institutions, but does not include garbage. Combustible rubbish (the "trash" component of rubbish) consists of paper, rags, cartons, boxes, wood, furniture, tree branches, yard trimmings, and so on. Some cities have separate designations for yard wastes. Combustible rubbish is not putrescible and may be stored for long periods of time. Noncombustible rubbish is material that cannot be burned at ordinary incinerator temperatures of 700 to 1 100 °C. It is the inorganic portion of refuse, such as tin cans, heavy metals, glass, ashes, and so on.

Table 5.1 Refuse materials by kind, composition, and sources

kind	Composition	Sources
Garbage	Wastes from preparation, cooking, and serving of food; market wastes; wastes from handling, storage, and sale of produce	Households, restaurants, institutions, stores, markets
Rubbish	Combustible: paper, cartons, boxes, barrels, wood, excelsior, tree branches, yard trimmings, wood furniture, bedding, dunnage	
	Noncombustible: metals, tin cans, metal furniture, dirt, glass, crockery, minerals	
Ashes	Residue from fires used for cooking and heating and from on-site incineration	
Street refuse	Sweepings, dirt, leaves, catch basin dirt, contents of litter receptacles	Streets, sidewalks, alleys, vacant lots
Dead animals	Cats, dogs, squirrels, deer	
Abandoned vehicles	Unwanted cars and trucks left on public property	
Industrial wastes	Food-processing wastes, boiler house cinders, lumber scraps, metal scraps, shavings	Factories, power plants
Demolition wastes	Lumber, pipes, brick, masonry, and other construction materials from razed buildings and other structures	Demolition sites to be used for new buildings, renewal projects, expressways
Construction wastes	Scrap lumber, pipe, other construction materials	New construction, remodeling
Special wastes	Hazardous solids and liquids; explosives, pathological wastes, radioactive materials	Households, hotels, hospitals, institutions, stores, industry
Sewage treatment tesidue	Solids from coarse screening and from grit chambers; septic tank sludge	Sewage treatment plants, septic tanks

(Source: Mackenzie L. Davis and David A. Cornwell, *Introduction to Environmental Engineering*, Tsinghua University Press, 2007, P739)

Composition Refuse composition is possibly the most important characteristic affecting its disposal or the recovery of materials and energy from refuse. The chemical composition of typical municipal solid waste (MSW) is shown in Table 5.2. Composition may vary significantly from one community to the next and with time in any given community.

Table 5.2 Proximate and ultimate chemical analysis of MSW

Proximate analysis	
Moisture (%)	15-35
Volatile matter (%)	50-60
Fixed carbon (%)	3-9
Noncombustibles (%)	15-25
Higher heat value (Btu/lb)	3,000-6,000
Ultimate analysis	
Moisture (%)	15-35
Carbon (%)	15-30
Hydrogen (%)	2-5
Oxygen (%)	12-24
Nitrogen (%)	0.2-1.0
Sulfur (%)	0.02-0.1

(Source: Mackenzie L. Davis and David A. Cornwell, *Introduction to Environmental Engineering*, Tsinghua University Press, 2007, P739)

Refuse composition is expressed in terms of either "as generated" or "as disposed" since during the disposal process moisture transfer takes place, thus changing the weights of the various fractions. Table 5.3 shows typical breakdowns of average refuse components for the United States. It should be re-emphasized that such numbers are useful only as guidelines and that each community has unique characteristics that influence its solid waste production and composition.

Table 5.3 Average composition of MSW in the United States

Category	As Generated		As Disposed	
	Millions of tons	%	Millions of tons	%
Paper	37.2	29.2	44.9	35.3
Glass	13.3	10.2	13.5	10.6
Metal	10.1	7.9	10.1	7.9
Ferrous	8.8	6.9	8.8	6.9

(Source: P. Aarne Vesilind J. Jeffrey Peirce and Ruth F. Weiner, *Environmental Engineering*, Butterworths, 1988, P252)

Table 5.3 Average composition of MSW in the United States

Category	As Generated		As Disposed	
	Millions of tons	%	Millions of tons	%
Aluminum	0.9	0.7	0.9	0.7
Other nonferrous	0.4	0.3	0.4	0.3
Plastics	6.4	5.0	6.4	5.0
Rubber and leather	2.6	2.0	3.4	2.7
Textiles	2.1	1.6	2.2	1.7
Wood	4.9	3.8	4.9	3.8
Food waste	22.8	17.9	19.1	15.0
Yard waste	26.0	20.4	20.0	15.7
Miscellaneous	1.9	1.5	2.8	2.1
Total	127.3		127.3	

(Source: P. Aarne Vesilind J. Jeffrey Peirce and Ruth F. Weiner, *Environmental Engineering*, Butterworths, 1988, P253)

5.2 DISPOSAL

The disposal of solid wastes is defined as placement of the waste so it no longer impacts society. This is achieved either by assimilating the residue so it can no longer be identified in the environment (e.g., fly ash from an incinerator) or by hiding the wastes well enough so they cannot be readily found.

The common methods of solid-waste disposal include on-site disposal, composting, incineration, and sanitary landfills. Of these, the physical factors are most significant in siting of sanitary landfills. Without careful consideration of the soils, rocks, and hydrogeology, a landfill program may not function properly.

5.2.1 On-site Disposal

By far the most common on-site disposal method in urban areas is the mechanical grinding of kitchen food waste. Garbage disposal devices are installed in the waste-water pipe system from a kitchen sink, and the garbage is ground and flushed into the sewer system. This effectively reduces the amount of handling and quickly removes food waste, but final disposal is transferred to the sewage treatment plant where solids such as sewage sludge still must be disposed of. Hazardous liquid chemicals may also be inadvertently or deliberately disposed of in sewers, requiring treatment plants to handle toxic materials. Illegal dumping in urban sewers has been identified as a potential major problem.

Another method of on-site disposal is small-scale incineration. This method is common in institutions and apartment houses. It requires constant attention and periodic maintenance to insure proper operation. In addition, the ash and other residue must be removed periodically and transported to a final disposal site.

5.2.2 Composting

Composting is the aerobic decomposition of organic matter by microorganisms, primarily bacteria and fungi. The reactions generate heat, raising compost temperatures during the composting period. Waste volume is reduced by about 30 percent for wastes with a high proportion of newsprint to perhaps 60 percent for garden debris.

Composting may take place naturally under controlled conditions or in mechanized composting plants. In natural systems, ground garbage, preferably with glass and metals removed, is mixed with a nutrient source (sewage sludge, animal manure, night soil) and a filler (wood chips, ground corn cobs) which permits air to enter the pile. The mixture, maintained about 50 percent moisture content, is placed in windrows, 2 to 3 m wide, and turned over once or twice a week. In four to six weeks, when the color darkens, the temperature drops, and a musty odor develops, the process is complete. The filler may then be removed and the remaining "humus" used as soil conditioner. With mechanical plants, continual aeration and mixing enable composting time to be reduced by about 50 percent. A short period usually follows the mechanical process to allow the composting material to "mature".

Although composting is not common in the United States, it is popular in Europe and Asia, where many successful solid waste composting plants have been operating for many years. For example, Rotterdam, in Holland, has a major composting plant to complement its waste management program. The same interest in composting exists in developing countries, but in these areas windrow systems are the preferred method.

5.2.3 Incineration

Incineration is the reduction of combustible waste to inert residue by burning at high temperatures (800 °C-1 000 °C). Generally, the process of incineration includes solid waste storage, pretreatment, feeding system, a combustion chamber, exhaust emissions, pollution control, slagging, monitoring, testing, and energy recovery etc.. Figure 5.1 shows a schematic illustration of a typical incinerator for refuse.

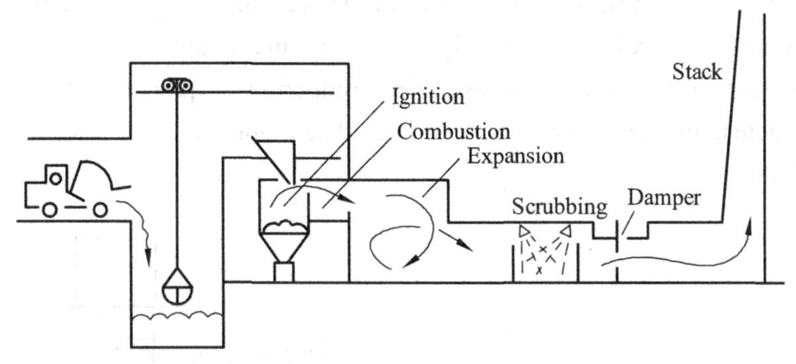

Figure 5.1 A typical incinerator for refuse

(Source: P. Aarne Vesilind, *Environmental Pollution And Control*, Ann Arbor Science, 1975, P104)

Incineration for reducing waste volume by about 90 percent and weight by 75 percent with the possibility of energy recovery, became a very popular processing option in many developed countries. Incineration of waste materials converts the waste into ash, flue gas, and heat. The ash is mostly formed by the inorganic constituents of the waste, and may take the form of solid lumps or particulates carried by the flue gas. The flue gases must be cleaned of gaseous and particulate pollutants before they are dispersed into the atmosphere. In some cases, the heat generated by incineration can be used to generate electric power. Incineration with energy recovery is one of several waste-to-energy (WtE) technologies such as gasification, pyrolysis, and anaerobic digestion. Denmark and Sweden have been leaders in using the energy generated from incineration for more than a century, in localised combined heat and power facilities supporting district heating schemes. In 2005, waste incineration produced 4.8% of the electricity consumption and 13.7% of the total domestic heat consumption in Denmark. A number of other European countries rely heavily on incineration for handling municipal waste, in particular Luxembourg, the Netherlands, Germany and France.

Incineration has particularly strong benefits for the treatment of certain waste types in niche areas such as clinical wastes and certain hazardous wastes where pathogens and toxins can be destroyed by high temperatures.

However, in several countries, there are still concerns from experts and local communities about the environmental impact of incinerators. Incinerators built just a few decades ago often did not include a materials separation to remove hazardous, bulky or recyclable materials before combustion. These facilities tended to risk the health of plant workers and local environment due to inadequate levels of gas cleaning and combustion process control. Most of these facilities did not generate electricity.

The newer municipal incinerators are usually the continuously burning type, and many have "waterwall" construction in the combustion chamber in place of the older, more common refractory lining. The waterwall consists of joined vertical boiler tubes containing water. The tubes absorb the heat to provide hot water for steam, and they also control the furnace temperature. With waterwall units, costly refractory maintenance is eliminated, pollution control requirements are reduced (because of the reduction in quench water and gas volumes requiring treatment), and heat recovery is simpler. The components of a waterwall incinerator are shown in Figure 5.2.

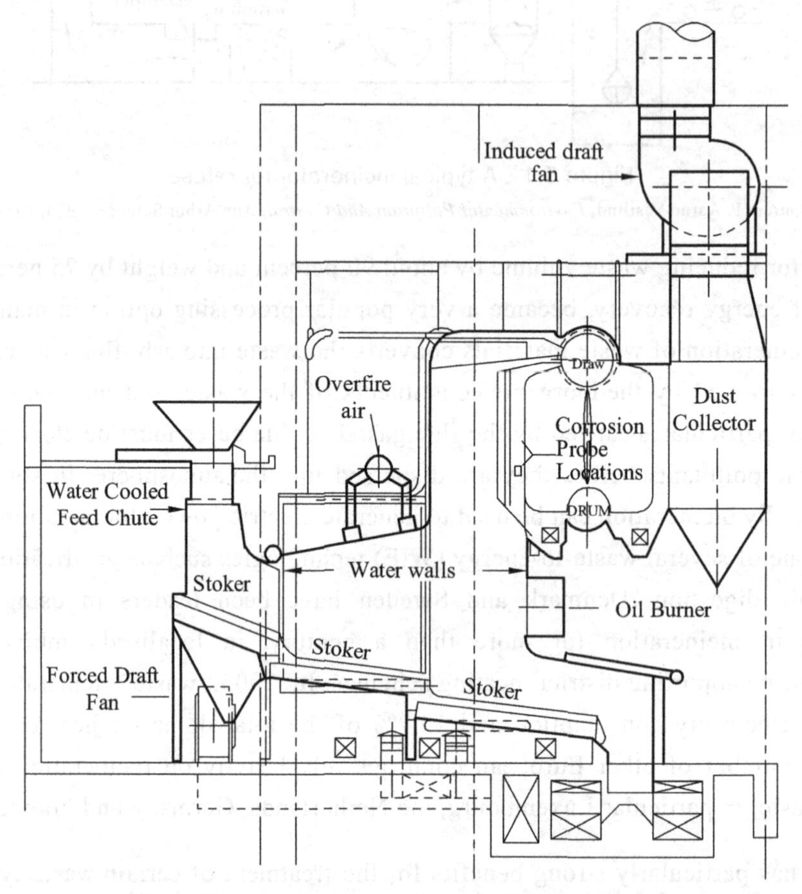

Figure 5.2 Section through a waterwall incinerator

(Source: J. Glynn Henry and Gary W. Heinke, *Environmental Science And Engineering*, Prentice Hall, 1989, P557)

The combustion temperatures of conventional incinerators fueled only by wastes are about 760 °C (1,400 °F) in the furnace proper and in excess of 870 °C (1,600 °F) in the secondary combustion chamber. These temperatures are needed to avoid odor from incomplete combustion. Temperatures up to 1,650 °C (3,000 °F), which would reduce volume by 97 percent and convert metal and glass to ash, are possible with supplementary fuels.

Of the problems associated with incineration, air pollution control, especially the removal of the fine particulates and toxic gases (including dioxin), is the most difficult. The emission of combustible, carbon-containing pollutants can be controlled by optimizing the combustion process. Oxides of nitrogen and sulfur and other gaseous pollutants have not been a problem because of their relatively small concentration. Other concerns related to incineration include the disposal of the liquid wastes from floor drainage, quench water, and scrubber effluent, and the problem of ash disposal in landfills because of heavy metal residues. However, in a study from 1997, Delaware Solid Waste Authority found that, for the same amount of produced energy, incineration plants emitted fewer particles, hydrocarbons and less SO_2, HCl, CO and NO_x than coal-fired power plants, but more than natural gas-fired power plants. According to Germany's Ministry of the Environment, waste incinerators reduce the amount of some atmospheric pollutants by substituting power produced by coal-fired plants with power from waste-fired plants.

5.2.4 Sanitary Landfill

Except for the disposal of municipal solid wastes at sea, which is not permitted by most developed countries, solid wastes, or their residues in some form, must go to the land. From the earliest times, disposal of solid wastes into open dumps was standard practice for municipalities. The town dump was usually in a low-lying area near a watercourse. Fires, water pollution, odors, rats, flies, and blowing papers were the visible results. Burial of the waste reduced these problems. A ***sanitary landfill*** as defined by the American Society of Civil Engineering is a method of solid waste disposal that functions without creating a nuisance or hazard to public health or safety. Engineering principles are used to confine the waste to the smallest practical area, reduce it to the smallest practical volume, and cover it with a layer of compacted soil at the end of each day of operation, or more frequently if necessary. This covering of the waste with compacted soils makes the sanitary landfill "sanitary". The compacted layer effectively denies continued access to the waste by insects, rodents, and other animals. It also isolates the refuse from the air, thus minimizing the amount of surface water entering into and gas escaping from the waste.

The sanitary landfill, presumably to distinguish it from the typical unsanitary open dump, was first used in California in 1934 to reclaim land. Process design, leachate collection, gas treatment and site monitoring, together with more care in site selection, are some of the key processing technologies of sanitary landfill.

Site Selection A landfill site is a site for the disposal of waste materials by burial. Factors controlling the feasibility of sanitary landfills include topographic relief, location of the groundwater table, amount of precipitation, type of soil and rock, and the location of the disposal zone in the surface-water and ground-water flow system. The best site are those in which natural conditions ensure reasonable safety in disposal of solid waste; conditions may be safe because of climatic, hydrologic, or geologic conditions, or combinations of these. Ideally, a sanitary landfill site should be on inexpensive land within economical hauling distance, have year-round access, and be at least 1,500 m downwind from residential and commercial neighbors. The area should be reasonably clear, level, and well drained, with capacity for not less than about three years' use until its future role as "open" space is realized. Soil of low permeability, well above the groundwater table, is also desirable for protection of underground water supplies and as cover material. Final choice of the site should not be made without a detailed hydrogeological investigation.

Preparation of the site involves fencing, grading, stockpiling of cover material, construction of berms, landscaping, and the installation of leachate collection and monitoring systems. Wells for gas collection may also be provided.

Mixed wastes with varying degrees of compaction are delivered to the site in packer trucks and/or trailer units. Some hand sorting of incoming wastes will be necessary, and pulverizing or high-pressure compaction and baling for volume reduction may precede placement. Loose material is placed in the lower part of the prepared pit or trench and then spread and compacted by machine in layers of about 0.5 m thickness. When the depth reaches 2 to 3 m, and at the end of each day's operation, the refuse is covered with 150 to 300 mm of earth. This consolidated solid waste enclosed by earth is called a cell and normally contains one day's waste. The cross section of the sanitary landfill in Figure 5.3 shows its salient design features.

Landfilling mixed refuse of low density (300 kg/m^3) may be an uneconomical use of a site with limited capacity. Reduction of waste volume can not only extend the life of the landfill, but can contribute other benefits as well. Volume reduction by incineration is too costly, except for larger cities, so for small communities physical methods are the only practical alternative. Milling or pulverizing the refuse by a hammer mill is the most common method. In addition to a volume reduction of 50 percent and the reduction or perhaps elimination of the cover material, problems with odors, blowing papers, rodents, insects, fires, settlement, and mired vehicles are all greatly reduced.

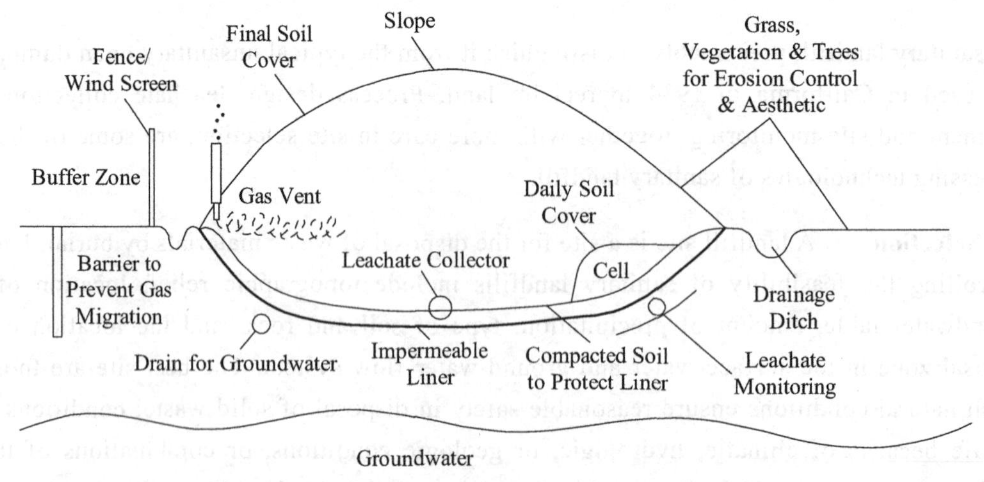

Figure 5.3 Cross section through a sanitary landfill

(Source: J. Glynn Henry and Gary W. Heinke, *Environmental Science And Engineering*, Prentice Hall, 1989, P563)

High-pressure compaction of municipal wastes into solid bales or blocks (about 1 m^3) weighing 850 to 950 kg is another way to reduce waste volume. The advantages are the same as those for pulverizing, but in addition, the need for on-site sorting and field compaction is eliminated. Furthermore, the resulting stable bales are resistant to infiltration of rainwater and, if necessary, suitable for rail haul. Because of the saving in landfill capacity and other advantages, greater use of pulverizing or high-pressure compaction for waste volume reduction seems likely in the future.

Operational Methods Although various titles are used to describe the operating methods employed at sanitary landfills, only two basic techniques are involved. They are termed as the area method (Figure 5.4) and the trench method (Figure 5.5). At many sites, both methods are used, either simultaneously or sequentially.

In the area method, the solid waste is deposited on the surface, compacted, then covered with a layer of compacted soil at the end of the working day. Use of the area method is seldom restricted by topography; flat or rolling terrain, canyons, and other types of depressions are all acceptable. The cover material may come from on or off site.

The trench method is used on level or gently sloping land where the water table is low. In this method a trench is excavated; the solid waste is placed in it and compacted; and the soil that was taken from the trench is then laid on the waste and compacted. The advantage of the trench method is that cover material is readily available as a result of trench excavation. Stockpiles can be created by excavating long trenches, or the material can be dug up daily. The depth depends on the location of the groundwater and/or the character of the soil. Trenches should be at least twice as wide as the compacting equipment so that the treads or wheels can compact all the material on the working area.

A MSW landfill does not need to be operated by using only the area or trench method.

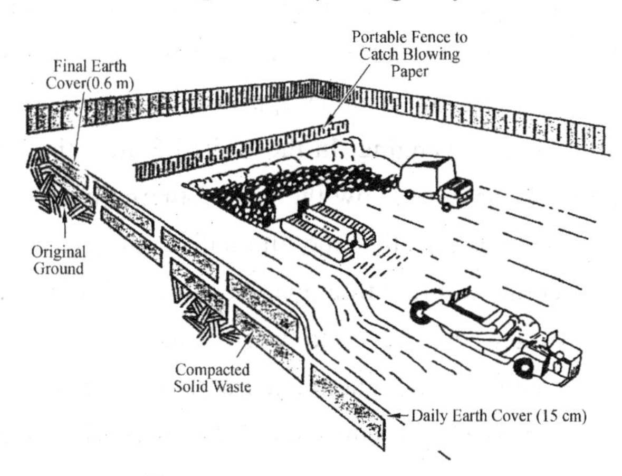

Figure 5.4 The area method

(Source: Mackenzie L. Davis and David A. Cornwell, *Introduction to Environmental Engineering*, Tsinghua University Press, 2007, P766

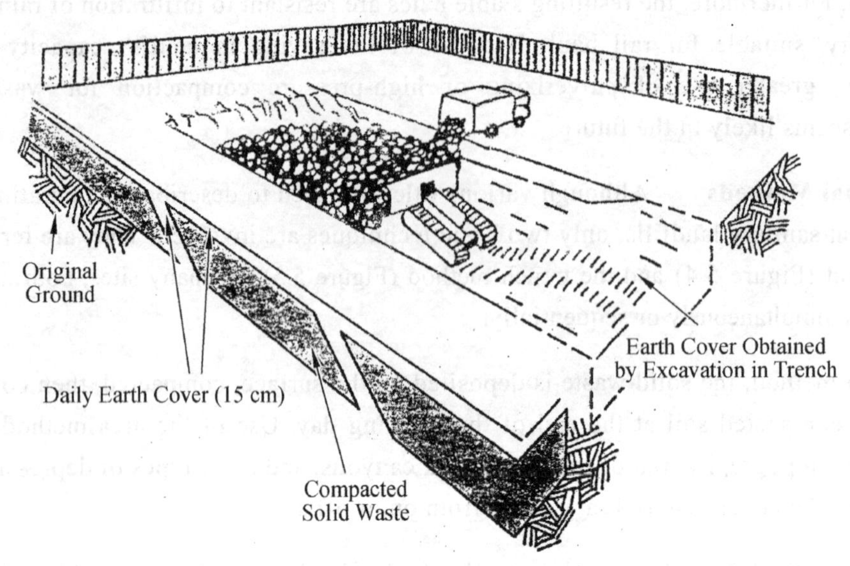

Figure 5.5 The trench method

(Source: Mackenzie L. Davis and David A. Cornwell, *Introduction to Environmental Engineering*, Tsinghua University Press, 2007, P767)

Combinations of the two are possible. The methods used can be varied according to the constraints of the particular site.

A profile view of a typical landfill is shown in Figure 5.6. The waste and the daily cover placed in a landfill during one operational period form a cell. The operational period is usually one day. The waste is dumped by the collection and transfer vehicles onto the working face. It is spread in 0.4 to 0.6 m layers and compacted by driving a crawler tractor or other compaction equipment over it. At the end of each day cover material is placed over the cell. The cover material may be native soil or other approved materials. Its purpose is to prevent fires, odors, blowing litter, and scavenging. In the United States, the federal regulations also permit the state regulatory authority to allow the use of alternative daily covers (ADC) if the owner of the landfill can demonstrate that the alternative material functions as well as the earthen cover without presenting a threat to human health or the environment. Some landfills have successfully demonstrated that diverted wastes such as chipped tires, yard waste, shredded wood waste, and petroleum-contaminated soils can be used effectively as ADCs. Using these waste products as ADCs presents a cost savings for the landfill and also increases the landfill's available space.

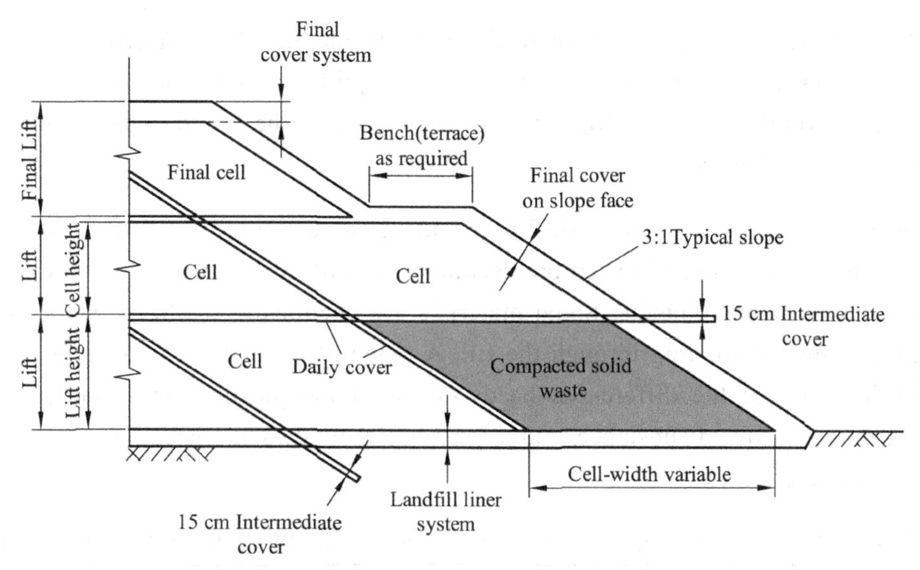

Figure 5.6 Sectional View Through a MSW Landfill

(Source: Mackenzie L. Davis and David A. Cornwell, *Introduction to Environmental Engineering*, Tsinghua University Press, 2007, P768)

Recommended depths of cover for various exposure periods are given in table 5.4. The dimensions of a cell are determined by the amount of waste and the operational period.

Table 5.4 Recommended depths of cover

Type of cover	Minimum depth(m)	Exposure time(d)
Daily	0.15	<7
Intermediate	0.30	7 to 365
Final	0.60	>365

(Source: Mackenzie L. Davis and David A. Cornwell, *Introduction to Environmental Engineering*, Tsinghua University Press, 2007, P768)

A lift may refer to the placement of a layer of waste or the completion of the horizontal active area of the landfill. In Figure 5.6 a lift is shown as the completion of the active area of the landfill. An extra layer of intermediate cover may be provided if the lift is exposed for long periods. The active area may be up to 300 m in length and width. The side slopes typically range from 1.5 : 1 to 2 : 1. Trenches vary in length from 30 to 300 m with widths of 5 to 15 m. The trench depth may be 3 to 9 m.

Benches are used where the height of the landfill exceeds 15 to 20 m. They are used to maintain the slope stability of the landfill, for the placement of surface water drainage channels, and for the location of landfill gas collection piping.

Final cover is applied to the entire landfill site after all landfilling operations are complete. A modern final cover will contain several different layers of material to perform different functions.

Leachate Liquid that passes through the landfill and that has extracted dissolved and suspended matter from it is called leachate. The liquid enters the landfill from external sources such as rainfall, surface drainage, groundwater, and the liquid in and produced from the decomposition of the waste.

Solid wastes placed in a sanitary landfill may undergo a number of biological, chemical, and physical changes. Aerobic and anaerobic decomposition of the organic matter results in both gaseous and liquid end products. Some materials are chemically oxidized. Some solids are dissolved in water percolating through the fill. A range of leachate compositions is listed in Table 5.5. Because of the differential heads (slope of the piezometric surface), the water containing those dissolved substances moves into the groundwater system. The result is gross pollution of the groundwater.

Table 5.5 Typical data of the composition of leachate from new and mature landfill

Constituent	Value(mg/L)		
	New landfill (less than 2 years)		Mature landfill (greater than 10 years)
	Range	Typical	
BOD_5 (5-day biochemical oxygen demand)	2,000-30,000	10,000	100-200
TOC (total organic carbon)	1,500-20,000	6,000	80-160
COD(chemical oxygen demand)	3,000-60,000	18,000	100-500
Total suspended solids	200-2,000	500	100-400
Organic nitrogen	10-800	200	80-120
Ammonia nitrogen	10-800	200	20-40
Nitrate	5-40	25	5-10
Total phosphorus	5-100	30	5-10
Ortho phosphorus	4-80	20	4-8
Alkalinity as $CaCO_3$	1,000-10,000	3,000	200-1,000
pH(no units)	4.5-7.5	6	6.6-7.5
Total hardness as $CaCO_3$	300-10,000	3,500	200-500
Calcium	200-3,000	1,000	100-400
Magnesium	50-1,500	250	50-200
Potassium	200-1,000	300	50-400
Sodium	200-2,500	500	100-200
Chloride	200-3,000	500	100-400
Sulfate	50-1,000	300	20-50
Total iron	50-1,200	60	20-200

(Source: Mackenzie L. Davis and David A. Cornwell, *Introduction to Environmental Engineering*, Tsinghua University Press, 2007, P773)

Proper planning, site selection, and operating normally can minimize the possibility of surface and groundwater pollution. Some common preventive measures are:

1. Locating the site at a safe distance from streams, lakes. and wells;

2. Avoiding site locations above porous soil;

3. Using an earth cover that is nearly impervious;

4. Providing suitable drainage.

In general, it has been found that the quantity of leachate is a direct function of the amount of external water entering the landfill. The rate of seepage of leachate from the bottom of a landfill can be estimated from Darcy's law.

Landfill Gases The principal gaseous products emitted from a landfill (methane and carbon dioxide) are the result of microbial decomposition. Typical concentrations of landfill gases and their characteristics are summarized in Table 5-6. During the early life of the landfill, the predominant gas is carbon dioxide. As the landfill matures, the gas is composed almost equally of carbon dioxide and methane. Because the methane is explosive, its movement must be controlled. The heat content of this landfill gas mixture (10,000 to 20,000 kJ/m^3), although not as substantial as methane alone (37,000 kJ/m^3), has sufficient economic value that many landfills have been tapped with wells to collect it.

Increasing fuel prices in the 1970s provoked interest in the possibility of gas recovery from sanitary landfills. At the end of 2004,there were 378 landfill gas (LFG) recovery projects in the United States, and most of them were under way to recover and purify landfill gas for on-site use for heat and power, or for off-site use as fuel. Methane (CH_4) constitutes 40 to 60 percent of the landfill gas. Theoretically, the total amount of gas produced is 200 to 270 L of CH_4 per kg of refuse, depending on the characteristics of the solid wastes and the basis of the determination. Of the amount generated, an estimated 15 to 35 percent can be recovered.

Table 5.6 Typical constituents found in MSW landfill gas

Component	Percent (dry volume basis)
Methane	45-60
Carbon dioxide	40-60
Nitrogen	2-5
Oxygen	0.1-1.0
Sulfides, disulfides, mercaptans, etc.	0-1.0
Ammonia	0.1-1.0
Hydrogen	0-0.2

Carbon monoxide	0-0.2
Trace constituents	0.01-0.06
Characteristic	Value
Temperature,°C	35-50
Specific gravity	1.02-1.05
Moisture content	Saturated
High heating value, kJ/m^3	16,000-20,000

(Source: Mackenzie L. Davis and David A. Cornwell, *Introduction to Environmental Engineering*, Tsinghua University Press, 2007, P769)

Landfill stabilization and, hence, gas generation take place over a long time. Thirty years is a commonly mentioned period, but this could be shortened under continuously wet conditions or prolonged if the refuse remained dry. For a stabilization period of 25 to 30 years, one-third to two-thirds of the gas might be generated within the first five years.

Because of their toxicity, trace gas emissions from landfills are of concern. More than 150 compounds have been measured at various landfills. Many of these may be classified as volatile organic compounds (VOCs). The occurrence of significant VOC concentrations is often associated with older landfills that previously accepted industrial and commercial wastes containing these compounds.

Completed sanitary landfills Completed landfills generally require maintenance because of uneven settling. Maintenance consists primarily of regrading the surface to maintain good drainage and filling in small depressions to prevent ponding and possible subsequent groundwater pollution. The final soil cover should be about 0.6 m deep.

Completed landfills have been used for recreational purposes such as parks, playgrounds, or golf courses. Parking and storage areas or botanical gardens are other final uses. Because of the characteristic uneven settling and gas evolution from landfills, construction of buildings on completed landfills should be avoided.

On occasion, one-story buildings and runways for light aircraft might be constructed. In such cases. it is important to avoid concentrated foundation loading, which can result in uneven settling and cracking of the structure. The designer must provide the means for the gas to dissipate into the atmosphere and not into the structure.

Problems with landfilling Many of the shortcomings of a poorly operated landfill are evident: odors and blowing papers carried by the wind; vermin, insects, and scavenger birds attracted by the organic refuse; and dust and noise from trucks and compacting operations. Continuous field compaction of the loose refuse and covering all material with earth at the end of each day alleviates these problems. Volume reduction by pulverizing or high-pressure compaction provides even greater assurance of an aesthetically acceptable operation.

Property devoted to landfilling is no longer available as productive farm land or as taxable property. Even after closure of the site, future use of the area must be restricted to some type of open development such as a park, recreational area, or ski hill, and the construction of buildings must be rigidly controlled.

Under warm, moist conditions, organic wastes become ideal breeding places for disease-causing organisms. Pathogens, even if absent initially, have easy access to the waste via vectors. With solid waste, the usual vectors for disease transmission, i.e., water, air, and food, are not important; flies, rodents, and mosquitoes are the primary vectors. The preventative measures for public health include:

1. The use of tightly closed containers for organic wastes.
2. Compaction of waste to at least 600 kg/m^3 to reduce insect breeding places and rodent access.
3. Processing within two days (since fly larvae become flies in a few days).
4. Shredding of waste to promote aerobic decomposition.

Contamination of the groundwater by leachate high in organics, dissolved solids, and other constituents can be a serious problem where nearby wells are used for water supply. The hazard stems mainly from soluble salts, since biodegradable organics and pathogenic microorganisms are usually removed by the soil before the leachate has traveled very far.

The gases, principally methane (CH_4) and carbon dioxide (CO_2) generated by the anaerobic decomposition of organics in the landfill are also a concern. Methane is an odorless, combustible gas that is heavier than air and explosive when its concentration in air is between 5 and 15 percent. It is therefore a hazardous gas and should not be ignored. Carbon dioxide in combination with water creates an acidic environment in which minerals such as calcium, magnesium iron, cadmium, lead, and zinc that are present in the refuse (or in the soil) tend to dissolve and move toward the groundwater table. Calcium and magnesium only add hardness to groundwater, but the toxic heavy metals are a more serious problem because they can make the water unfit for human consumption.

Chapter 6
Noise Pollution

With the technological expansion of the Industrial Revolution and continuous development through a post-World War II acceleration, environmental noise in the industrialized nations has been gradually and steadily increasing, with more geographic areas becoming exposed to significant levels of noise. Once noise level is sufficient to some intensity and duration, it may adversely affect us and other biological species.

In this chapter we discuss first the basics of sound, defining some of the terms that acoustical engineers use to describe and control unwanted sound (noise). We then discuss some of the effects of noise, ending with a discussion on noise control.

6.1 SOUND

6.1.1 Concept of Noise

Sound is in effect a transfer of energy. For example, rocks thrown at you would certainly get your attention, but this would require the transfer of mass (rocks). Alternatively, your attention may be gained by poking you with a stick, in which case the stick is not lost, but energy is transferred from the poker to the pokee. In the same way, sound travels through a medium such as air without a transfer of mass. Just as the stick has to move back and forth, so must air molecules oscillate in waves to transfer energy.

Noise, commonly defined as unwanted sound, is an environmental phenomenon to which we are exposed before birth and throughout life. Noise can also be considered an environmental pollutant, a waste product generated in conjunction with various anthropogenic activities. Under the latter definition, ***noise*** is any sound – independent of loudness – that can produce an undesired physiological or psychological effect in an individual, and that may interfere with the social ends of an individual or group. These social ends include all of our activities – communication, work, rest, recreation, and sleep.

There are valid reasons why widespread recognition of noise as a significant environmental pollutant and potential hazard or, as a minimum, a detractor from the quality of life has been slow in coming. In the first place, noise, if defined as unwanted sound, is a subjective experience. What is considered noise by one listener may be considered desirable by another.

Secondly, noise has a short decay time and thus does not remain in the environment for extended periods, as do air and water pollution. By the time the average individual is spurred to action to abate, control, or, at least, complain about sporadic environmental noise, the noise may no longer exist.

Thirdly, the physiological and psychological effects of noise on us are often subtle and insidious, appearing so gradually that it becomes difficult to associate cause with effect. Indeed, to those persons whose hearing may already have been affected by noise, it may not be considered a problem at all.

Further, the typical citizen is proud of this nation's technological progress and is generally happy with the things that technology delivers, such as rapid transportation, labor-saving devices, and new recreational devices. Unfortunately, many technological advances have been associated with increased environmental noise, and large segments of the population have tended to accept the additional noise as part of the price of progress.

6.1.2 Properties of Sound Waves

Noise is a type of sound; therefore, it possesses all acoustic characteristics and rules of sound. Sound waves result from the vibration of solid objects or the separation of fluids as they pass over, around, or through holes in solid objects. The vibration and/or separation causes the surrounding air to undergo alternating compression and rarefaction, much in the same manner as a piston vibrating in a tube (Figure 6.1). The compression of the air molecules causes a local increase in air density and pressure. Conversely, the rarefaction causes a local decrease in density and pressure. These alternating pressure changes are the sound detected by the human ear.

If the piston vibrates at a constant rate, the condensations and rarefactions will move down the tube at a constant speed. That speed is the speed of sound (c). The rise and fall of pressure at point A will follow a cyclic or wave pattern over a "period" of time (Figure 6.2). The wave pattern is called ***sinusoidal***. The time between successive peaks or between successive troughs of the oscillation is called the ***period*** (P). The inverse of this, that is, the number of times a peak arrives in one second of oscillations, is called the ***frequency*** (f). Period and frequency are then related as follows:

$$P = \frac{1}{f} \tag{6-1}$$

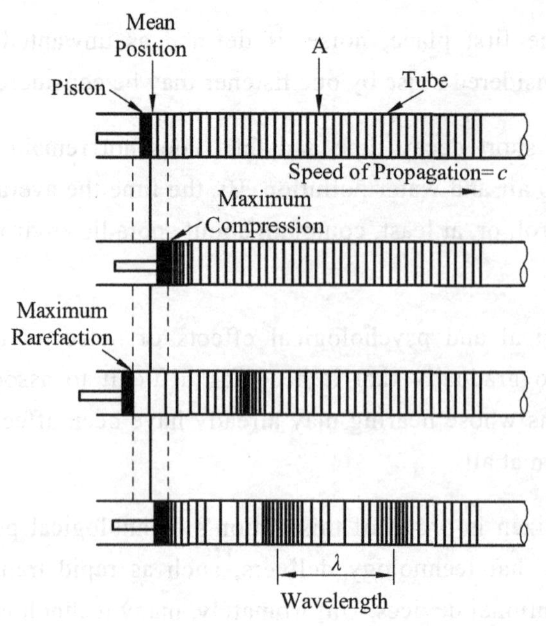

Figure 6.1 Alternating compression and rarefaction of air molecules resulting from a vibrating piston

(Source: Mackenzie L. Davis and David A. Cornwell, *Introduction to Environmental Engineering*, Tsinghua University Press, 2007, P655)

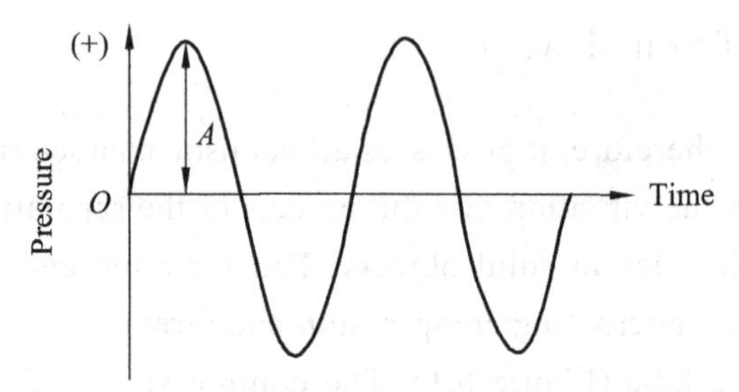

Figure 6.2 Sinusoidal wave that results from alternating compression and rarefaction of air molecules

(Source: Mackenzie L. Davis and David A. Cornwell, *Introduction to Environmental Engineering*, Tsinghua University Press, 2007, P656)

Since the pressure wave moves down the tube at a constant speed, you would find that the distance between equal pressure readings would remain constant. The distance between adjacent crests or troughs of pressure is called the ***wavelength*** (λ). Wavelength and frequency are then related as follows:

$$\lambda = \frac{c}{f} \tag{6-2}$$

where λ = wavelength, m

c = velocity of the sound in a given medium, m/sec

f = frequency, cycles/sec

Sound travels at different speeds in different materials, depending on the material's elasticity.

In acoustics, the frequency as cycles per second is denoted by the name hertz, and written as Hz. The common audible range for humans is between 20 and 20,000 Hz. The frequency is one of the two basic parameters that describe sound. Amplitude is the other.

The ***amplitude*** (A) of the wave is the height of the peak or depth of the trough measured from the zero pressure line (Figure 6.2). From Figure 6.2 we can also note that the average pressure could be zero if an averaging time was selected that corresponded to the period of the wave. This would result regardless of the amplitude! This, of course, is not an acceptable state of affairs. The root mean square (rms) sound pressure (p_{rms}) is used to overcome this difficulty. The rms sound pressure is obtained by squaring the value of the amplitude at each instant in time; summing the squared values: dividing the total by the averaging time; and taking the square root of the total. The equation for *rms* is

$$P_{rms} = (\overline{P^2})^{1/2} = \left[\frac{1}{T}\int_0^T P^2(t)\mathrm{d}t\right]^{1/2} \tag{6-3}$$

where the overbar rears to the time weighted average and T is the time period of the measurement.

6.1.3 Sound Power and Intensity

Work is defined as the product of the magnitude of the displacement of a body and the component of force in the direction of the displacement. Thus, traveling waves of sound pressure transmit energy in the direction of propagation of the wave. The rate at which this work is done is defined as the sound power (W).

Sound intensity (I) is defined as the time-weighted average sound power per unit area normal to the direction of propagation of the sound wave. Intensity and power are related as follows:

$$I = \frac{W}{A} \tag{6-4}$$

where A is a unit area perpendicular to the direction of wave motion. Intensity, and hence, sound power, is related to sound pressure in the following manner:

$$I = \frac{(P_{rms})^2}{\rho c} \tag{6-5}$$

where

I = intensity, W/m^2

P_{rms}= root mean square sound pressure, Pa

ρ = density of medium, kg/m^3

c = speed of sound in medium, m/s

Both the density of air and speed of sound are a function of temperature. The speed of sound in air at 101,325 kPa may be determined from the following equation:

$$c = 20.05\sqrt{T} \tag{6-6}$$

where T is the absolute temperature in °K and c is in m/s.

6.1.4 Levels and Decibel

The sound pressure of the faintest sound that a normal healthy individual can hear is about 0.00002 Pa. The sound pressure produced by a Saturn rocket at liftoff is greater than 200 Pa. Even in scientific notation this is an "astronomical" range of numbers.

In order to cope with this problem, a scale based on the logarithm of the ratios of the measured quantities is used. Measurements on this scale are called ***levels***. The unit for these types of measurement scales is the bel, which was named after Alexander Graham Bell:

$$L' = \lg\frac{Q}{Q_0} \tag{6-7}$$

where L'= level, bels

Q = measured quantity

Q_0 = reference quantity

lg = logarithm in base 10

A bel turns out to be a rather large unit, so for convenience it is divided into 10 subunits called decibels (dB). Levels in dB are computed as follows:

$$L = 10\lg\frac{Q}{Q_0} \tag{6-8}$$

The dB does not represent any physical unit. It merely indicates that a logarithmic transformation has been performed.

Sound power level If the reference quantity (Q_0) is specified, then the dB takes on physical significance. For noise measurements, the reference power level has been established as 10^{-12} watts. Thus, sound power level may be expressed as

$$L_W = 10\lg\frac{W}{10^{-12}} \tag{6-9}$$

Sound power levels computed with Equation 6-9 are reported as dB re: 10^{-12} W.

Sound intensity level The reference quantity is 10^{-12} W/m^2. Thus, the sound intensity level is given as

$$L_I = 10\lg\frac{I}{10^{-12}} \tag{6-10}$$

Sound pressure level Because sound measuring instruments measure the root mean square pressure, the sound pressure level is computed as follows:

$$L_p = 10\lg\frac{(p_{\mathrm{rms}})^2}{(p_{\mathrm{rms}})_0^2} \tag{6-11}$$

which, after extraction of the squaring term, is given as

$$L_p = 20\lg\frac{p_{\mathrm{rms}}}{(p_{\mathrm{rms}})_0} \tag{6-12}$$

The reference pressure has been established as 20 micropascals (μPa). A scale showing some common sound pressure levels is shown in Figure 6.3.

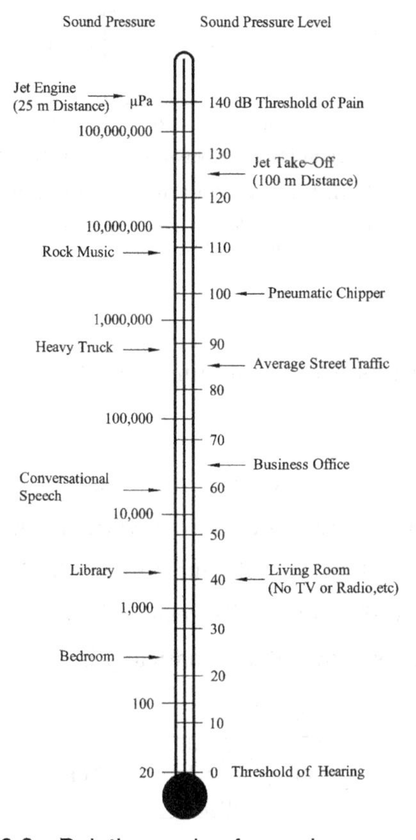

Figure 6.3 Relative scale of sound pressure levels

(Source: Mackenzie L. Davis and David A. Cornwell, *Introduction to Environmental Engineering*, Tsinghua University Press, 2007, P658)

Combining sound pressure levels Because of their logarithmic heritage, decibels don't add and subtract the way apples and oranges do. If you take a 60 decibel noise (re: 20 μPa) and add another 60 decibel noise (re: 20 μPa) to it, you get a 63 decibel noise (re: 20 μPa). For skeptics, this can be demonstrated by converting the dB to relative powers, adding them, and converting back to dB. Figure 6.4 provides a graphical solution to this type of problem. For noise pollution work, results should be reported to the nearest whole number. When there are several levels to be combined they should be combined two at a time, starting with lower-valued levels and continuing two at a time with each successive pair until one number remains. Henceforth in this chapter we will assume levels are all "re: 20 μPa" unless stated otherwise.

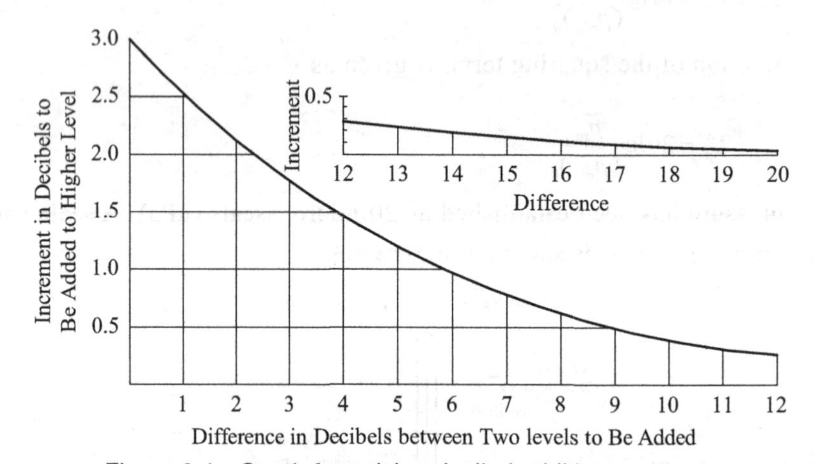

Figure 6.4 Graph for solving decibel addition problems

(Source: Mackenzie L. Davis and David A. Cornwell, *Introduction to Environmental Engineering*, Tsinghua University Press, 2007, P659)

Example 6.1

What sound pressure level results from combining the following three levels (all re: 20 μPa): 68 dB, 79 dB, 75 dB?

We begin by selecting the two lowest levels: 68 dB and 75 dB. The difference between the values is 75 – 68=7.00. Using Figure 6.4 to draw a vertical line from 7.00 on the abscissa to intersect the curve. A horizontal line from the intersection to the ordinate yields about 0.8 dB. Thus, the combination of 68 dB and 75 dB results in a level of 75.8 dB. This, and the remainder of the computation, is shown diagramatically below.

68 dB
Δ=7
75 dB
75.8 dB
Δ=3.2
79 dB
80.7 dB

Rounding off to the nearest whole number yields an answer of 81 dB re: 20 μPa.

6.1.5 Loudness

In general, two pure tones have different frequencies but the same sound pressure level will be heard as different loudness levels. Loudness level is a psychoacoustic quantity.

In 1933, Fletcher and Munson conducted a series of experiments to determine the relationship between frequency and loudness. A reference tone and a test tone were presented alternately to the test subjects. They were asked to adjust the sound level of the test tone until it sounded as loud as the reference. The results were plotted as sound pressure level in dB versus the test tone frequency (Figure 6.5). The curves are called the Fletcher-Munson or equal loudness contours. The reference frequency is 1,000 Hz. The curves are labeled in phons, which are the sound pressure levels of the 1,000 Hz pure tone in dB. The lowest contour (dashed line) represents the "threshold of hearing". The actual threshold may vary by as much as ±10 dB between individuals with normal hearing.

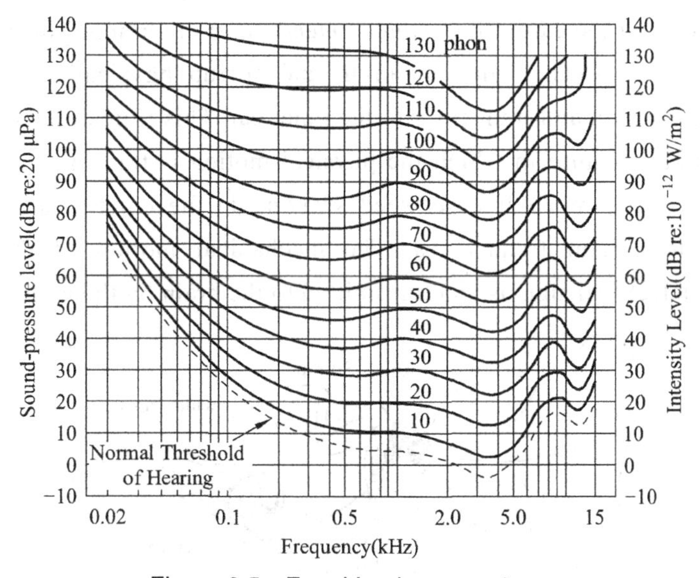

Figure 6.5 Equal loudness contours

(Source: Mackenzie L. Davis and David A. Cornwell, *Introduction to Environmental Engineering*, Tsinghua University Press, 2007, P669)

6.2 EFFECTS OF NOISE

6.2.1 Health Effects of Noise

The health effects of noise on people are classified into the following two categories: auditory effects and psychological/sociological effects. Auditory effects include both hearing loss and speech interference. Psychological/sociological effects include annoyance, sleep interference, effects on performance, and acoustical privacy.

In the Bronx borough of New York City, one spring evening, four boys were at play, shouting and racing in and out of an apartment building. Suddenly, from a second-floor window, came the crack of a pistol. One of the boys sprawled dead on the pavement. The victim happened to be thirteen years old, son of a prominent public leader, but there was no political implication in the tragedy. The killer confessed to police that he was a nightworker who had lost control of himself because the noise from the boys prevented him from sleeping.

We only recently become aware of the devastating psychological effects of noise. The effect of excessive noise on our ability to hear, on the other hand, has been known for a long time.

The human ear is an incredible instrument. Imagine having to design and construct a scale for weighing just as accurately as a flea or an elephant. Yet this is the range of performance to which we are accustomed from our ears.

A schematic of the human auditory system is shown in Figure 6.6. Sound pressure waves caused by vibrations set the eardrum (*tympanic membrane*) in motion. This activates the three bones in the middle ear. The hammer, anvil, and stirrup physically amplify the motion received from the eardrum and transmit it to the inner ear. This fluid-filled cavity contains the cochlea, a snail-like structure in which the physical motion is transmitted to tiny hair cells. These hair cells deflect, much like seaweed swaying in the current, and certain cells are responsive only to certain frequencies. The mechanical motion of these hair cells is transformed to bioelectrical signals and transmitted to the brain by the auditory nerves.

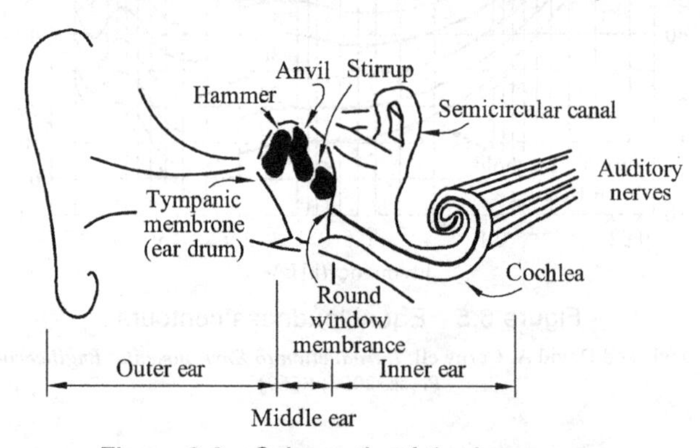

Figure 6.6 Schematic of the human ear

(Source: P. Aarne Vesilind, *Environmental Pollution And Control*, Ann Arbor Science, 1975, P185)

Acute damage may occur to the eardrum, but this occurs only with very loud sudden noises. More serious is the chronic damage to the tiny hair cells in the inner ear. Prolonged exposure to noise of a certain frequency pattern may cause either temporary hearing loss, which disappears in a few hours or days, or permanent loss. The former is called *temporary threshold shift* and the latter is known as *permanent threshold shift*. Literally, your threshold of hearing changes, so you are not able to hear some sounds.

Temporary threshold shift is generally not damaging to your ear unless it is prolonged. People who work in noisy environments commonly find that they hear less well at the end of the day. Performers in rock bands are subjected to very loud noises (substantially above the allowable OSHA levels, OSHA represents that noise is regulated by federal or national legislation called the Occupational Safety and Health Act in the industrial environment.) and commonly are victims of temporary threshold shift.

Repeated noise over a long time leads to permanent threshold shift. This is especially true in industrial applications in which people are subjected to noises of a certain frequency. Figure 6.7 shows data from a study performed on workers at a textile mill. Note that people who worked in the spinning and weaving parts of the mill, where noise levels are highest, suffered the most severe loss in hearing, especially at around 4,000 Hz, the frequency of noise emitted by the machines.

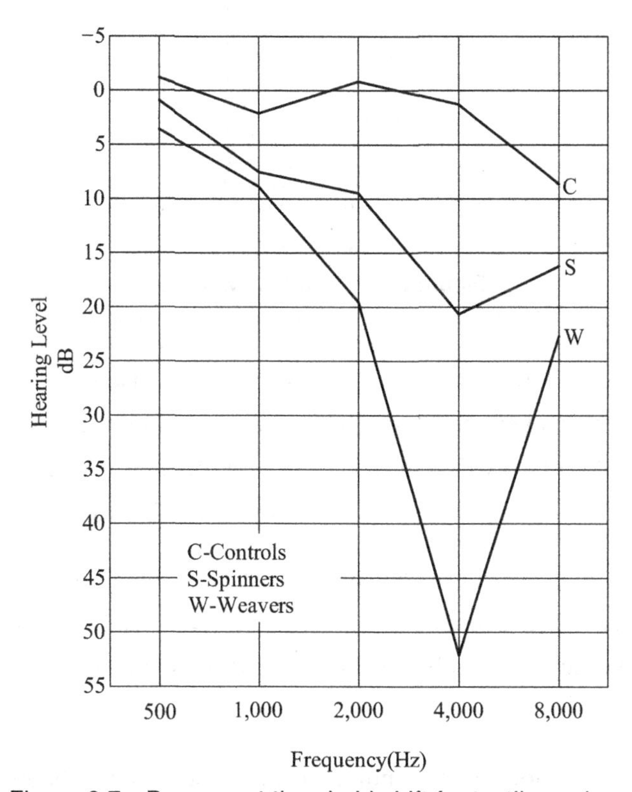

Figure 6.7 Permanent threshold shift for textile workers

(Source: P. Aarne Vesilind J. Jeffrey Peirce and Ruth F. Weiner, *Environmental Engineering*, Butterworths, 1988, P474)

As people get older, hearing becomes less acute simply as one of the effects of aging. This loss of hearing, called *presbycusis*, is illustrated in Figure 6.8. Note that the greatest loss occurs at the higher frequencies. Speech frequency is about 1,000 to 2,000 Hz, and thus older people commonly accuse others of "whispering."

In addition to presbycusis, however, there is a serious loss of hearing owing to environmental noise. In one study, 11 percent of ninth graders, 13 percent of twelfth graders, and 35 percent of college freshmen had a greater than 15 dB loss of hearing at 2,000 Hz. The study concluded that this severe loss resulted from exposure to loud noises such as motorcycles and rock music and that as a result "the hearing of many of these students had already deteriorated to a level of the average 65-year-old person".

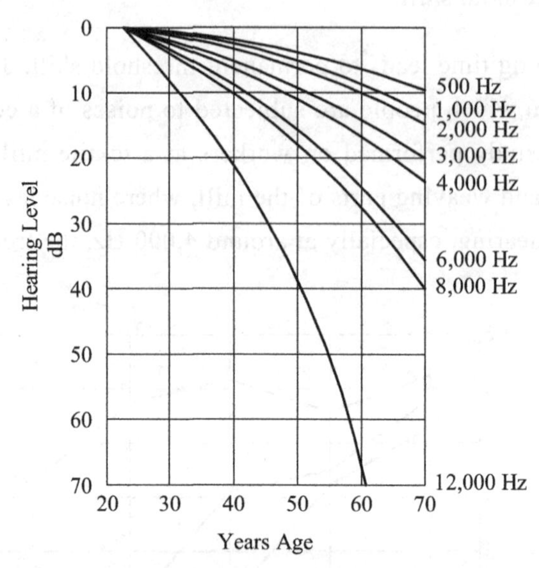

Figure 6.8 Hear loss with age

(Source: P. Aarne Vesilind J. Jeffrey Peirce and Ruth F. Weiner, *Environmental Engineering*, Butterworths, 1988, P475)

Noise also affects other bodily functions such as the cardiovascular system. Noise alters the rhythm of the heartbeat, makes the blood thicker, dilates blood vessels, and makes focusing difficult. It is no wonder that excessive noise has been blamed for headaches and irritability. Noise is especially annoying to people who do close work, like watchmakers.

All of the above reactions are those that our ancestral cavemen also experienced. Noise to them meant danger, and their senses and nerves were "up", ready to repel the danger. In the modern noise-filled world, we are always "up", and it is unknown how much of our physical ills are due to noise.

We also know that we cannot adapt to noise in the sense that our body functions no longer reactin a certain way to excessive noise. People do not, therefore, "get used to" noise in the physiological sense.

In addition to the noise problem, it might be appropriate to mention the potential problems of very high or very low frequency sound, out of our usual 20 to 20,000 Hz hearing range. The health effects of these, if any, remain to be studied.

6.2.2 The Dollar Cost of Noise

Numerous case histories comparing patients in noisy and quiet hospitals point to increased convalescence time when the hospital was noisy (either from within or owing to external noise). This may be translated directly to a dollar figure.

Recent court cases have been won by workers seeking damages for hearing loss suffered during work. The Veterans Administration spends many, many millions of dollars every year for care of patients with hearing disorders.

Other costs, such as sleeping pills, lost time in industry, and apartment soundproofing are difficult to quantify. The John F. Kennedy Cultural Center in Washington spent $5 million for soundproofing, necessitated by the jets using the nearby National Airport.

6.2.3 Degradation of environmental Quality

It is even more difficult to measure the effect noise has had on the quality of life. How much is noise to blame for irate husbands and grumpy wives, for grouchy taxi drivers and surly clerks?

Children reared in noisy neighborhoods must be taught to listen. They cannot focus their auditory senses on one sound, such as the voice of a teacher.

Within the 80-to 90-dB circle around the Kennedy airport in New York are 22 schools. Every time a plane passes the teacher must stop talking and try to re-establish attention.

The effect of these results cannot be measured. Yet we intuitively feel that the effect must be negative.

Noise is a real and dangerous form of environmental pollution. Since people cannot adapt to it physiologically, we are perhaps adapting psychologically instead. Noise may keep our senses "on edge" and prevent us from relaxing. Our mental powers must therefore control this result to our bodies. Since noise, in the context of human evolution, is a very recent development, we have not yet adapted to it, and must thus be living on our buffer capacity. One wonders how plentiful this is.

6.3 RATING SYSTEMS

6.3.1 Goals of a Noise-rating system

An ideal noise-rating system is one that allows measurements by sound level meters or analyzers to be summarized succinctly, and yet represents noise exposure in a meaningful way. In our previous discussions on loudness and annoyance, we noted that your response to sound is strongly dependent on the frequency of the sound. Furthermore, we noted that the type of noise (continuous, intermittent, or impulsive) and the time of day that it occurred (night being worse than day) were significant factors in annoyance.

Thus, the ideal system must take frequency into account. It should differentiate between daytime and nighttime noise. And, finally, it must be capable of describing the cumulative noise exposure. A statistical system can satisfy these requirements.

The practical difficulty with a statistical rating system is that it would yield a large set of parameters for each measuring location. A much larger array of numbers would be required to characterize a neighborhood. It is literally impossible for such an array of numbers to be used effectively in enforcement. Thus, there has been a considerable effort to define a single number measure of noise exposure.

6.3.2 L_N and L_{eq} Systems

The L_N concept The parameter L_N is a statistical measure that indicates how frequently a particular sound level is exceeded. If, for example, we write L_{40}=72 dBA, then we know that 72 dB(A) was exceeded for 40% of the measuring time. A plot of L_N against N where N=1%, 2%, 3%, and so forth, would look like the cumulative distribution curve shown in Figure 6.9.

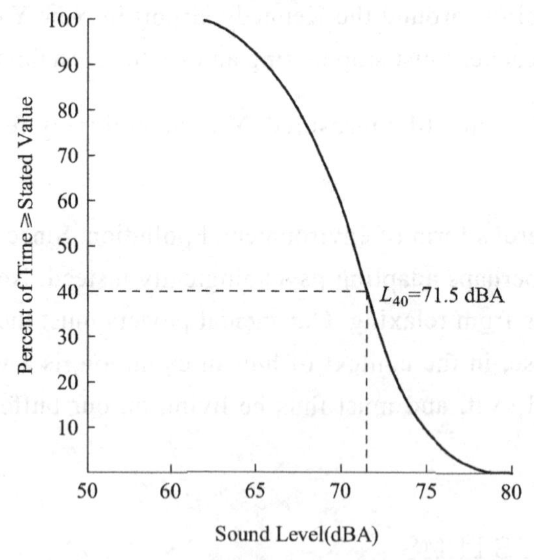

Figure 6.9 Cumulative distribution curve

(Source: Mackenzie L. Davis and David A. Cornwell, *Introduction to Environmental Engineering*, PWS Engineering, 1985, P430)

Allied to the cumulative distribution curve is the probability distribution curve. A plot of this will show how often the noise levels fall into certain class intervals. In Figure 6.10 we can see that 22% of the time the measured noise levels ranged between 70 and 72 dBA; for 17% of the time they ranged between 72 and 74 dBA and so on. The relationship between this picture and the one for L_N is really quite simple. By adding the percentages given in successive class intervals from right to left, we can arrive at a corresponding L_N where N is the sum of the percentages and L is the lower limit of the left most class interval added, thus, L_{40}:

$$L_{(2+7+14+17)} = 72 \text{ dBA}$$

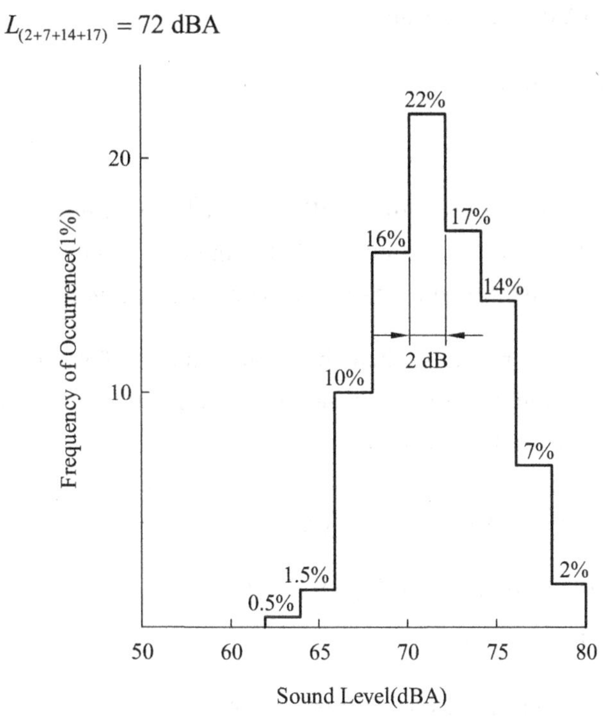

Figure 6.10 Probability distribution plot

(Source: Mackenzie L. Davis and David A. Cornwell, *Introduction to Environmental Engineering*, PWS Engineering, 1985, P430)

The L_{eq} concept The equivalent continuous equal energy level, (L_{eq}) can be applied to any fluctuating noise level. It is that constant noise level that, over a given time expends the same amount of energy as the fluctuating level over the same time period. It is expressed as follows:

$$L_{eq} = 10\lg\frac{1}{t}\int_0^t 10^{L(t)/10}\,dt \tag{6-13}$$

where

t = the time over which L_{eq} is determined

$L(t)$ = the time varying noise level in dBA

Generally speaking, there is no well-defined relationship between $L(t)$ and time so a series of discrete samples of $L(t)$ have to be taken. This modifies the expression to:

$$L_{eq} = 10\lg\sum_{i=1}^{i=n} 10^{L_i/10} t_i \qquad (6\text{-}14)$$

where

n = the total number of samples taken

L_i = the noise level in dBA of the i-th sample

t_i = fraction of total sample time

Example 6.2

Consider the case where a noise level of 90 dBA exists for 5 minutes and is followed by a reduced noise level of 60 dBA for 50 minutes, what is the equivalent continuous equal energy level for the 55-minute period? Assume a 5-minute sampling interval.

If the sampling interval is 5 minutes, then the total number of samples (n) is 11 and the fraction of total sample time (t_i) for each sample is 1/11=0.091. With these preliminary calculations, we may now compute the sum:

$$\sum_{i=1}^{i=n} = (10^{90/10})(0.091) + (10^{60/10})(0.91)$$

$$= (0.91\times10^7) + (9.1\times10^5) = 9.19\times10^7$$

And finally, we take the log to find

$$L_{eq} = 10\log(9.19\times10^7) = 79.6 \text{ or } 80 \text{ dBA}$$

The example calculation is depicted graphically in Figure 6.11. From this you may note that large emphasis is put on occasional high noise levels.

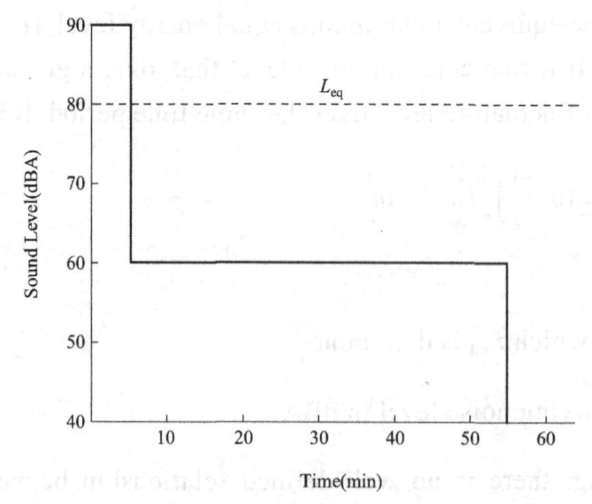

Figure 6.11 Graphical illustration of L_{eq} computation given in example 6.2

(Source: Mackenzie L. Davis and David A. Cornwell, *Introduction to Environmental Engineering*, PWS Engineering, 1985, P432)

The equivalent noise level was introduced in 1965 in Germany as a rating specifically to evaluate the impact of aircraft noise upon the neighbors of airports. It was almost immediately recognized in Austria as appropriate for evaluating the impact of street traffic noise in dwellings and schoolrooms. It has been embodied in the National Test Standards of both East Germany and West Germany for rating the subjective effects of fluctuating noises of all kinds, such as from street and road traffic, rail traffic, canal and river ship traffic, aircraft, industrial operations (including the noise from individual machines), sports stadiums, playgrounds and the like.

6.4 NOISE CONTROL

6.4.1 Source-path-receiver Concept

If you have a noise problem and want to solve it, you have to find out something about what the noise is doing, where it comes from, how it travels, and what can be done about it. A straightforward approach is to examine the problem in terms of its three basic elements: that is, sound arises from a source, travels over a path, and affects a receiver, or listener.

The source may be one or any number of mechanical devices that radiate noise or vibratory energy. Such a situation occurs when several appliances or machines are in operation at a given time in a home or office.

The most obvious transmission path by which noise travels is simply a direct line-of-sight air path between the source and the listener. For example, aircraft flyover noise reaches an observer on the ground by the direct line-of-sight air path. Noise also travels along structural paths. Noise can travel from one point to another via any one path or a combination of several paths. Noise from a washing machine operating in one apartment may be transmitted to another apartment along air passages such as open windows, doorways, corridors, or ductwork. Direct physical contact of the washing machine with the floor or walls sets these building components into vibration. This vibration is transmitted structurally throughout the building, causing walls in other areas to vibrate and to radiate noise.

The receiver may be a single person, a classroom of students, or a suburban community.

Solution of a given noise problem might require alteration or modification of any or all of these three basic elements:

1. modifying the source to reduce its noise output;

2. altering or controlling the transmission path and the environment to reduce noise level reaching the listener;

3. providing the receiver with personal protective equipment.

6.4.2 Control of Noise Source by Design

Reduce impact forces Many machines and items of equipment are designed with parts that strike forcefully against other parts, producing noise. Often, this striking action or impact is essential to the machine's function. A familiar example is the typewriter–its keys must strike the ribbon and paper in order to leave an inked impression. But the force of the key also produces noise as the impact falls on the ribbon, paper, and platen.

Several steps can be taken to reduce noise from impact forces. The particular remedy to be applied will be determined by the nature of the machine in question. Not all of the steps listed below are practical for every machine and for every impact-produced noise. But application of even one suggested measure can often reduce the noise appreciably.

Some of the more obvious design modifications are as follows:

1. Reduce the weight, size, or height of fall of the impacting mass.

2. Cushion the impact by inserting a layer of shock-absorbing material between the impacting surfaces. (For example, insert several sheets of paper in the typewriter behind the top sheet to absorb some of the noise-producing impact of the keys.) In some situations, you could insert a layer of shock-absorbing material behind each of the impacting heads or objects to reduce the transmission of impact energy to other parts of the machine.

3. Whenever practical, one of the impact heads or surfaces should be made of non-metallic material to reduce resonance (ringing) of the heads.

4. Substitute the application of a small impact force over a long time period for a large force over a short period to achieve the same result.

5. Smooth out acceleration of moving parts by applying accelerating forces gradually. Avoid high-jerky acceleration or jerky motion.

6. Minimize overshoot, backlash, and loose play in cams, followers, gears, linkages, and other parts. This can be achieved by reducing the operational speed of the machine, better adjustment, or by using spring-loaded restraints or guides. Machines that are well made, with parts machined to close tolerances, generally produce a minimum of such impact noise.

Reduce speeds and pressures Reducing the speed of rotating and moving parts in machines and mechanical systems results in smoother operation and lower noise output. Likewise, reducing pressure and flow velocities in air, gas, and liquid circulation systems lessens turbulence, resulting in decreased noise radiation. Some specific suggestions that may be incorporated in design are the following:

1. Fans, impellers, rotors, turbines, and blowers should be operated at the lowest bladetip speeds that will still meet job needs. Use large-diameter low-speed fans rather than small-diameter high-speed units for quiet operation. In short, maximize diameter and minimize tip speed.

2. All other factors being equal, centrifugal squirrel-cage type fans are less noisy than vane axial or propeller type fans.

3. In air ventilation systems, a 50 percent reduction in the speed of the air flow may lower the noise output by 10 to 20 dB, or roughly half, or a quarter of the original loudness. Air speeds less than 3 m/s are measured at a supply or return grille produce a level of noise that usually is unnoticeable in residential or office areas. In a given system, reduction of air speed can be achieved by operating at lower motor or blower speeds, installing a greater number of ventilating grilles, or increasing the cross-sectional area of the existing grilles.

Reduce frictional resistance Reducing friction between rotating, sliding, or moving parts in mechanical systems frequently results in smoother operation and lower noise output. Similarly, reducing flow resistance in fluid distribution systems results in less noise radiation.

Four of the more important factors that should be checked to reduce frictional resistance in moving parts are the following:

1. Alignment: Proper alignment of all rotating, moving, or contacting parts results in less noise output. Good axial and directional alignment in pulley systems, gear trains shaft coupling, power transmission systems, and bearing and axle alignment are fundamental requirements for low noise output.

2. Polish: Highly polished and smooth surfaces between sliding, meshing, or contacting parts are required for quiet operation, particularly where bearings, gears, cams, rails, and guides are concerned.

3. Balance: Static and dynamic balancing of rotating parts reduces frictional resistance and vibration, resulting in lower noise output.

4. Eccentricity (out-of-roundness): Off-centering of rotating parts such as pulleys, gears, rotors, and shaft/bearing alignment causes vibration and noise. Likewise, out-of-roundness of wheels, rollers and gears causes uneven wear, resulting in flat spots that generate vibration and noise.

The key to effective noise control in fluid systems is streamline flow. This holds true regardless of whether one is concerned with air flow in ducts or vacuum cleaners, or water flow in plumbing systems. Streamline flow is simply smooth, nonturbulent low-friction flow.

The two most important factors that determine whether flow will be streamline or turbulent are the speed of the fluid and the cross-sectional area of the flow path, that is, the pipe or duct diameter. The rule of thumb for quiet operation is to use a low-speed, large-diameter system to meet a specified flow capacity requirement. However, even such a system can inadvertently generate noise if certain aerodynamic design features are overlooked or ignored. A system designed for quiet operation will employ the following features:

1. Low fluid speed: Low fluid speeds avoid turbulence, which is one of the main causes of noise.

2. Smooth boundary surfaces: Duct or pipe systems with smooth interior walls, edges, and joints generate less turbulence and noise than systems with rough or jagged walls or joints.

3. Simple layout: A well-designed duct or pipe systems with a minimum of branches, turns, fittings, and connectors is substantially less noisy than a complicated layout.

4. Long-radius turns: Changes in flow direction should be made gradually and smoothly. It has been suggested that turns should be made with a curve radius equal to about five times the pipe diameter or major cross-sectional dimension of the duct.

5. Flared sections: Flaring of intake and exhaust openings, particularly in a duct system, tends to reduce flow speeds at these locations, often with substantial reductions in noise output.

6. Streamline transition in flow path: Changes in flow path dimensions or cross-sectional areas should be made gradually and smoothly with tapered or flared transition sections to avoid turbulence. A good rule of thumb is to keep the cross-sectional area of the flow path as large and as uniform as possible throughout the system.

7. Remove unnecessary obstacles: The greater the number of obstacles in the flow path, the more tortuous, turbulent, and hence noisier, the flow. All other required and functional devices in the path, such as structural supports, deflectors, and control dampers, should be made as small and streamlined as possible to smooth out the flow patterns.

Reduce radiating area Generally speaking, the larger the vibrating part or surface, the greater the noise output. The rule of thumb for quiet machine design is to minimize the effective radiating surface areas of the parts without impairing their operation or structural strength. This can be done by making parts smaller, removing excess material, or by cutting openings slots, or perforations in the parts. For example, replacing a large vibrating sheet metal safety guard on a machine with a guard made of wire mesh or metal webbing might result in a substantial reduction in noise, because of the drastic reduction in surface area of the part.

Reduce noise leakage In many cases, machine cabinets can be made into rather effective soundproof enclosures through simple design changes, and the application of some sound-absorbing treatment. Substantial reductions in noise output may be achieved by adopting some of the following recommendations:

1. All unnecessary holes or cracks, particularly at joints, should be caulked.

2. All electrical or plumbing penetrations of the housing or cabinet should be sealed with rubber gaskets or a suitable non-setting caulk.

3. If practical, all other functional or required openings or ports that radiate noise should be covered with lids or shields edged with soft rubber gaskets to effect an airtight seal.

4. Other openings required for exhaust, cooling, or ventilation purposes should be equipped with mufflers or acoustically lined ducts.

5. Openings should be directed away from the operator and other people.

Isolate and dampen vibrating elements In all but the simplest machines, the vibration energy from a specific moving part is transmitted through the machine structure, forcing other component parts and surfaces to vibrate and radiate sound – often with greater intensity than that generated by the originating source itself.

Generally, vibration problems can be considered in two parts. First, we must prevent energy transmission between the source and surfaces that radiate the energy. Second, we must dissipate or attenuate the energy somewhere in the structure. The first part of the problem is solved by isolation. The second part is solved by damping.

The most effective method of vibration isolation involves the resilient mounting of the vibrating component on the most massive and structurally rigid part of the machine. All attachments or connections to the vibrating part, in the form of pipes, conduits, and shaft couplers, must be made with flexible or resilient connectors or couplers. For example, pipe connections to a pump that is resiliently mounted on the structural frame of a machine should be made of resilient tubing and be mounted as close to the pump as possible. Resilient pipe supports or hangers may also be required to avoid bypassing the isolated system (see Figure 6.12).

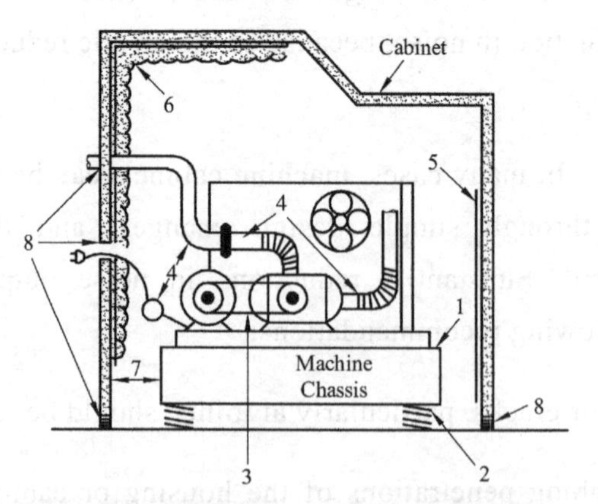

Figure 6.12 Examples of vibration isolation

1 – Motors, pumps, and fans installed on most massive part of the machine.
2 – Resilient mounts or vibration isolators used for the installation.
3 – Belt-drive or roller-drive systems used in place of gear trains.
4 – Flexible hoses and wiring used instead of rigid piping and stiff wiring.
5 – Vibration-damping materials applied to surfaces undergoing most vibration.
6 – Acoustical lining installed to reduce noise buildup inside machine.
7 – Mechanical contact minimized between the cabinet and the machine chassis.
8 – Openings at the base and other parts of the cabinet scaled to prevent noise leakage.

(Source: Mackenzie L. Davis and David A. Cornwell, *Introduction to Environmental Engineering*, Tsinghua University Press, 2007, P714)

Damping material or structures are those that have some viscous properties. They tend to bend or distort slightly, thus consuming part of the noise energy in molecular motion. The use of spring mounts on motors and laminated galvanized steel and plastic in air conditioning ducts are two examples.

When the vibrating noise source is not amenable to isolation, as, for example, in ventilation ducts, cabinet panels and covers, then damping materials can be used to reduce the noise.

The type of material, best suited for a particular vibration problem depends on a number of factors such as size, mass, vibrational frequency, and operational function of the vibrating structure. Generally speaking, the following guidelines should be observed in the selection and use of such materials to maximize vibration damping efficiency.

1. Damping materials should be applied to those sections of a vibrating surface where the most flexing, bending or motion occurs. These usually are the thinnest sections.

2. For a single layer of damping material, the stiffness and mass of the material should be comparable to that of the vibrating surface to which it is applied. This means that single-layer damping materials should be about two or three times as thick as the vibrating surface to which they are applied.

3. Sandwich materials (laminates) made up of metal sheets bonded to mastic (sheet metal viscoelastic composites) are much more effective vibration dampers than single-layer materials; the thickness of the sheet metal constraining layer and the viscoelastic layer should each be about one-third the thickness of the vibrating surface to which they are applied. Ducts and panels can be purchased already fabricated as laminates.

Provide mufflers/silencers There is no real distinction between mufflers and silencers. They are often used interchangeably. They are in effect acoustical filters and are used when fluid flow noise is to be reduced. The devices can be classified into two fundamental groups: ***absorptive mufflers*** and ***reactive mufflers***. An absorptive muffler is one whose noise reduction is determined mainly by the presence of fibrous or porous materials, which absorb the sound. A reactive muffler is one whose noise reduction is determined mainly by geometry. It is shaped to reflect or expand the sound waves with resultant self-destruction.

Although there are several terms used to describe the performance of mufflers, the most frequently used appears to be insertion loss (IL). Insertion loss is the difference between two sound pressure levels that are measured at the same point in space before and after a muffler has been inserted. Since each muffler IL is highly dependent on the manufacture's selection of materials and configuration, we will not present general IL prediction equations.

6.4.3 Noise Control in the Transmission Path

After you have tried all possible ways of controlling the noise at the source, your next line of defense is to set up devices in the transmission path to block or reduce the flow of sound energy before it reaches your ears. This can be done in several ways: 1) absorbing the sound along the path, 2) deflecting the sound in some other direction by placing a reflecting harrier in its path, 3) containing the sound by placing the source inside a sound-insulating box or enclosure.

Selection of the most effective technique will depend upon various factors, such as the size and type of source, intensity and frequency range of the noise, and the nature and type of environment.

Separation We can make use of the absorptive capacity of the atmosphere as well as divergence as a simple, economical method of reducing the noise level. Air absorbs high-frequency sounds more effectively than it absorbs low-frequency sounds. However, if enough distance is available, even low-frequency sounds will be absorbed appreciably.

If you can double your distance from a point source, you will have succeeded in lowering the sound pressure level by 6 dB. It takes about a 10 dB drop to halve the loudness. If you have to contend with a line source such as a railroad train, the noise level drops by only 3 dB for each doubling of distance from the source. The main reason for this lower rate of attenuation is that line sources radiate sound waves that are cylindrical in shape. The surface area of such waves only increases twofold for each doubling of distance from the source. However, when the distance from the train becomes comparable to its length, the noise level will begin to drop at a rate of 6 dB for each subsequent doubling of distance.

Indoors, the noise level generally drops only from 3 to 5 dB for each doubling of distance in the near vicinity of the source. However, further from the source, reductions of only 1 or 2 dB occur for each doubling of distance due to the reflections of sound off hard walls and ceiling surfaces.

Absorbing materials Noise, like light, will bounce from one hard surface to another. In noise control work, this is called reverberation. If a soft, spongy material is placed on the walls, floors, and ceiling, the reflected sound will be diffused and soaked up (absorbed). Sound-absorbing materials are rated either by their Sound-absorption coefficients (α_{SAB}) at 125, 250, 500, 1,000, 2,000, and 4,000 Hz or by a single number rafting called noise reduction coefficient (NRC). If a unit area of open window is assumed to transmit all and reflect none of the acoustical energy that reaches it, it is assumed to be 100% absorbent. This unit area of totally absorbent surface is called a "sabin." The absorptive properties of acoustical materials are then compared with this standard. The performance is expressed as a fraction or percentage of the sabin (α_{SAB}). The NRC is the average of the α_{SAB}s at 250, 500, 1,000, and 2,000 Hz rounded to the nearest multiple of 0.05. The NRC has no physical meaning. It is a useful means of comparing similar materials.

Sound-absorbing materials such as acoustic tile, carpets, and drapes placed on ceiling, floor, or wall surfaces can reduce the noise level in most rooms by about 5 to 10 dB for high-frequency sounds, but only by 2 or 3 dB for low frequency sound. Unfortunately, such treatment provides no protection to an operator of a noisy machine who is in the midst of the direct noise field. For greatest effectiveness, sound-absorbing materials should be installed as close to the noise source as possible.

If you have a small or limited amount of sound-absorbing material and wish to make the most effective use of it in a noisy room, the best place to put it is in the upper trihedral corners of the room, formed by the ceiling and two walls. Due to the process of reflection, the concentration of sound is greatest in the trihedral corners of a room. Additionally, the upper corner locations also protect the lightweight fragile materials from damage.

Because of their light weight and porous nature, acoustical materials are ineffectual in preventing the transmission of either airborne or structure-borne sound from one room to another. In other words, if you can hear people walking or talking in the room or apartment above, installing acoustical tile on your ceiling will not reduce the noise transmission.

Acoustical lining Noise transmitted through ducts, pipe chases, or electrical channels can be reduced effectively by lining the inside surfaces of such passage-ways with sound-absorbing materials. In typical duct installations, noise reductions of the order of 10 dB/m for an acoustical lining 2.5 cm thick are well within reason for high-frequency noise. A comparable degree of noise reduction for the lower frequency sounds is considerably more difficult to achieve because it usually requires at least a doubling of the thickness and/or length of acoustical treatment.

Barriers and panels Placing barriers, screens, or deflectors in the noise path can be an effective way of reducing noise transmission, provided that the barriers are large enough in size, and depending upon whether the noise is high-frequency or low-frequency. High-frequency noise is reduced more effectively than low-frequency noise.

The effectiveness of a barrier is dependent on its location, its height, and its length. Referring to Figure 6.13, we can see that the noise can follow five different paths.

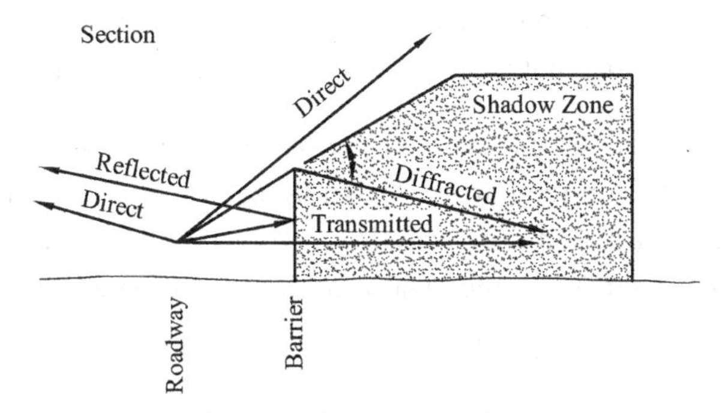

Figure 6.13 Noise paths from a source to a receiver

(Source: Mackenzie L. Davis and David A. Cornwell, *Introduction to Environmental Engineering*, Tsinghua University Press, 2007, P717)

First, the noise follows a direct path to receivers who can see the source well over the top of the barrier. The barrier does not block their line of sight (*L*/*S*) and therefore provides no attenuation. No matter how absorptive the barrier is, it cannot pull the sound downward and absorb it.

Second, the noise follows a diffracted path to receivers in the shadow zone of the barrier. The noise that passes just over the top edge of the barrier is diffracted (bent) down into the apparent shadow shown in the figure. The larger the angle of diffraction, the more the barrier attenuates the noise in this shadow zone. In other words, less energy is diffracted through large angles than through smaller angles.

Third, in the shadow zone, the noise transmitted directly through the barrier may be significant in some cases. For example, for extremely large angles of diffraction, the diffracted noise may be less than the transmitted noise. In this case, the transmitted noise compromises the performance of the barrier. It can be reduced by constructing a heavier barrier. The allowable amount of transmitted noise depends on the total barrier attenuation desired. Much information will begiven about this transmitted noise later.

The fourth path shown in Figure 6.13 is the reflected path. After reflection, the noise is of concern only to a receiver on the opposite side of the source. For this reason, acoustical absorption on the face of the barrier may sometimes be considered to reduce this reflected noise; however, this treatment will not benefit any receivers in the shadow zone. It should be noted that in most practical cases the reflected noise does not play an important role in barrier design. If the source of noise is represented by a line of noise, another short-circuit path is possible. Part of the source may be unshielded by the barrier. For example, the receiver might see the source beyond the ends of the barrier if the barrier is not long enough. This noise from around the ends may compromise, or short-circuit, barrier attenuation. The required barrier length depends on the total net attenuation desired. When 10 to 15 dB attenuation is desired, barriers must in general, be very long. Therefore, to be effective, barriers must not only break the line of sight to the nearest section of the source, but also to the source far up and down the line.

Of these four paths, the noise diffracted over the barrier into the shadow zone represents the most important parameter from the barrier design point of view. Generally, the determination of barrier attenuation or barrier noise reduction involves only calculation of the amount of energy diffracted into the shadow zone. The procedures presented in the barrier monograph used to predict highway noise are based on this concept.

Another general principle of barrier noise reduction that is worth reviewing at this point is the relation between noise attenuation expressed in decibels, energy terms, and subjective loudness. Table 6.1 gives these relationships for line sources. As indicated in the loudness column, a barrier attenuation of 3 dB will be barely discerned by the receiver. However, to attain this reduction, 50 percent of the acoustical energy must be removed. To cut the loudness of the source in half, a reduction of 10 dB is necessary. That is equivalent to eliminate 90 percent of the energy initially directed toward the receiver. As indicated previously, this drastic reduction in energy requires very long and high barriers. In summary, when designing barriers, you can expect the complexity of the design to be as follows:

Attenuation (dB)	Complexity
5	Simple
10	Attainable
15	Very difficult
20	Nearly impossible

Table 6.1 Relation between sound level reduction, energy, and loudness for line sources

To reduce A-level by dB	Remove portion of energy (%)	Divide loudness by
3	50	1.2
6	75	1.5
10	90	2
20	90	4
30	99.9	8
40	99.99	16

(Source: Mackenzie L. Davis and David A. Cornwell, *Introduction to Environmental Engineering*, Tsinghua University Press, 2007, P718)

Roadside barriers can be designed using the barrier nomograph in reverse order. A set of typical solutions is summarized in Table 6.2. The noise reduction at 152 m is less than that at 30 m because the barrier does not cast as large a shadow at a distance. The effectiveness of the barrier is reduced for trucks because of the elevated nature of the source.

Table 6.2 Noise reductions for various highway configurations

Highway configuration*		Height or depth (m)	Truck mix (%)	Noise reduction** at distance from ROW (dBA)	
Sketch	Description			30 m	152 m
	Roadside barriers 7.6 m from edge of shoulders; ROW = 78 m wide	6.1	0	13.9	13.3
			5	13.0	12.1
			10	12.6	11.7
			20	12.3	11.3
	Depressed roadway w/2 : 1 slopes; ROW = 102 m	6.1	0	9.9	11.4
			5	8.8	10.3
			10	8.4	9.8
			20	8.1	9.4
	Fill elevated roadway w/2 : 1 slopes; ROW = 102 m	6.1	0	9.0	6.3
			5	7.6	2.7
			10	7.1	1.8
			20	6.7	1.1
	Elevated structure; ROW = 78 m	7.3	0	9.8	6.0
			5	9.6	2.4
			10	9.3	1.5
			20	8.8	0.8

* Assumes divided 8 lanes with 9.1 m median.

** Based on observed 1.5 m above grade.

(Source: Mackenzie L. Davis and David A. Cornwell, *Introduction to Environmental Engineering*, Tsinghua University Press, 2007, P719)

Transmission loss When the position of the noise source is very close to the barrier, the diffracted noise is less important than the transmitted noise. If the barrier is in fact a wall panel that is sealed at the edges, the transmitted noise is the only one of concern.

The ratio of the sound energy incident on one surface of a panel to the energy radiated from the opposite surface is called the sound transmission loss (TL). The actual energy loss is partially reflected and partially absorbed. Since TL is frequency-dependent, only a complete octave or 1/3 octave band curve provides a full description of the performance of the barrier.

Enclosures Sometimes it is much more practical and economical to enclose a noisy machine in a separate room or box than to quiet it by altering its design, operation, or component parts. The walls of the enclosure should be massive and airtight to contain the sound. Absorbent lining on the interior surfaces of the enclosure will reduce the reverberant buildup of noise within it. Structural contact between the noise source and the enclosure must be avoided, or else the source vibration will be transmitted to the enclosure walls and thus short-circuit the isolation.

6.4.4 Control of Noise Source by Redress

The best way to solve noise problems is to design them out of the source. However, we are frequently faced with an existing source that, either because of age, abuse, or poor design, is a noise problem. The result is that we must redress, or correct the problem as it currently exists. The following sections identify some measures that might apply if you are allowed to tinker with the source.

Balance rotating parts One of the main sources of machinery noise is structural vibration caused by the rotation of poorly balanced parts, such as fans, fly wheels, pulleys, cams, shafts, and so on. Measures used to correct this condition involve the addition of counterweights to the rotating unit or the removal of some weight from the unit. You are probably familiar with noise caused by imbalance in the high-speed spin cycle of washing machines. The imbalance results from clothes not being distributed evenly in the tub. By redistributing the clothes, balance is achieved and the noise ceases. This same principle of balance can be applied to furnace fans and other common sources of such noise.

Reduce frictional resistance A well-designed machine that has been poorly maintained can become a serious source of noise. General cleaning and lubrication of all rotating, sliding, or meshing parts at contact points should go a long way toward fixing the problem.

Apply damping materials Since a vibrating body or surface radiates noise, the application of any material that reduces or restrains the vibrational motion of that body will decrease its noise output. Three basic types of redress vibration damping materials are available:

1. Liquid mastics, which are applied with a spray gun and harden into relatively solid materials, the most common being automobile "undercoating",

2. Pads of rubber, felt, plastic foam, leaded vinyls, adhesive tapes, or fibrous blankets, which are glued to the vibrating surface,

3. Sheet metal viscoelastic laminates or composites, which are bonded to the vibrating surface.

Seal noise leaks Small holes in an otherwise noise tight structure can reduce the effectiveness of the noise control measures. As you can see in Figure 6.14, if the designed transmission loss of an acoustical enclosure is 40 dB, an opening that comprises only 0.1% of the surface area will reduce the effectiveness of the enclosure by 10 dB.

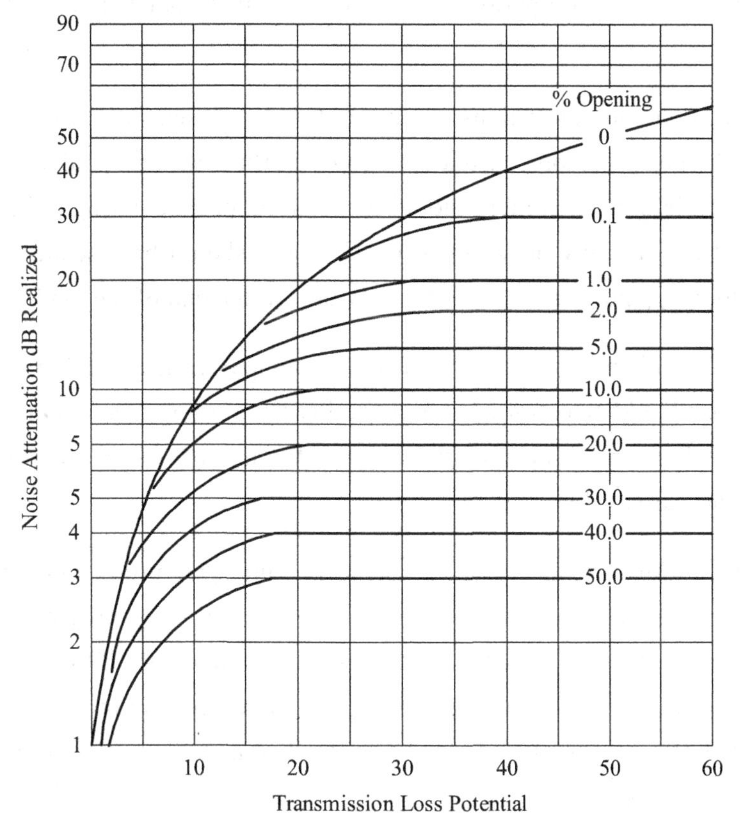

Figure 6.14 Transmission loss potential versus transmission loss realized for various opening sizes as a percent of total wall area

(Source: Mackenzie L. Davis and David A. Cornwell, *Introduction to Environmental Engineering*, Tsinghua University Press, 2007, P721)

Perform routine maintenance We all recognize the noise of a worn muffler. Likewise, studies of automobile tire noise in relation to pavement roughness show that maintenance of the pavement surface is essential to keep noise at minimum levels. Normal road wear can yield noise increases on the order of 6 dBA.

6.4.5 Protecting the Receiver

When all else fails When exposure to intense noise fields is required and none of the measures discussed so far is practical, as, for example, for the operator of a chain saw or pavement breaker, then measures must be taken to protect the receiver. The following two techniques are commonly employed.

Alter work schedule Limit the amount of continuous exposure to high noise levels. In terms of hearing protection, it is preferable to schedule an intensely noisy operation for a short interval of time each day over a period of several days rather than a continuous 8-hour run for a day or two.

In industrial or construction operations, an intermittent work schedule would benefit not only the operator of the noisy equipment, but also other workers in the vicinity. If an intermittent schedule is not possible, then workers should be given relief time during the day. They should take their relief time at a low noise level location, and should be discouraged from trading relief time for dollars, paid vacation, or an "early out" at the end of the day!

Inherently noisy operations, such as street repair, municipal trash collection, factory operation, and aircraft traffic, should be curtailed at night and early morning to avoid disturbing the sleep of the community. Remember: operations between 10 p.m. and 7 a.m. are effectively 10 dBA higher than the measured value.

Ear protection Molded and pliable earplugs, cup type protectors, and helmets are commercially available as hearing protectors. Such devices may provide noise reductions ranging from 15 to 35 dB (Figure 6.15). Earplugs are effective only if they are properly fitted by medical personnel. As shown in Figure 6.15, maximum protection can be obtained when both plugs and muffs are employed. Only muffs that have a certification stipulating the attenuation should be used.

These devices should be used only as a last resort after all other methods have failed to lower the noise level to acceptable limits. Examples include lawn mowing, mulchers, chippers, and weapon firing at target ranges. It should be noted that protective ear devices do interfere with speech communication and can be a hazard in some situations where warning calls may be a routine part of the operation (for example, TIMBERRRR!). A modern ear-destructive device is a portable mini-radio/recorder that uses earphones. In this "reverse" muff, high noise levels are directed at the ear without attenuation. If you can hear someone else's radio/recorder, that person is subjecting him-or her-self to noise levels in excess of 90-95 dBA!

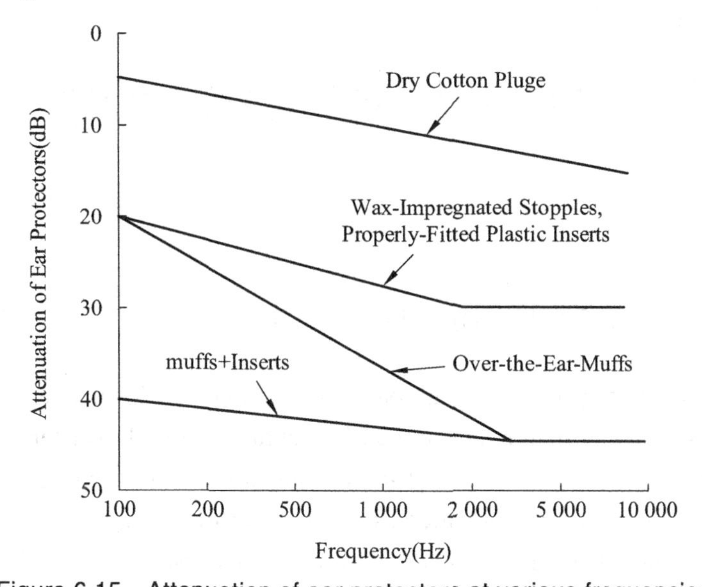

Figure 6.15 Attenuation of ear protectors at various frequencies

(Source: Mackenzie L. Davis and David A. Cornwell, *Introduction to Environmental Engineering*, Tsinghua University Press, 2007, P722)

Chapter 7
Climate Change and Global Warming

7.1 TERMINOLOGIES/DEFINITIONS

Climate change, illustrated mainly by global warming and sea level rise, is one of the most serious challenges facing human being in the 21st century. Natural disasters and other extreme climate phenomena are more often observed all around the world, whereas average temperature and global sea level are increasing at an unprecedented rate, becoming a major concern of all nations. The most general definition of *climate change* is a change in the statistical properties of the climate system when considered over long periods of time, regardless of cause. Accordingly, fluctuations over periods shorter than a few decades, such as El Niño, do not represent climate change.

The term sometimes is used to refer specifically to climate change caused by human activity, as opposed to changes in climate that may have resulted as part of the Earth's natural processes. In this sense, especially in the context of environmental policy, the term climate change has become synonymous with anthropogenic global warming. In scientific journals, global warming refers to surface temperature increases while climate change includes global warming and everything else that increasing greenhouse gas levels will affect.

An increase in the average temperature of the Earth's atmosphere, especially a sustained increase, is great enough to cause changes in the global climate. The Earth has experienced numerous episodes of global warming through its history, and currently appears to be undergoing another one. The present warming is generally attributed to an increase in the greenhouse effect , brought about by increased levels of greenhouse gases, largely due to the effects of human industry and agriculture. Expected long-term effects of current global warming are rising sea levels, flooding, melting of polar ice caps and glaciers, fluctuations in temperature and precipitation, more frequent and stronger El Niños and La Niñas, drought, heat waves, and forest fires.

Climate change: A change in the state of the *climate* that can be identified by changes in the mean and/or the variability of its properties and that persists for an extended period, typically decades or longer. Climate change may be due to natural internal processes or *external forcing*, or to persistent *anthropogenic* changes in the composition of the *atmosphere* or in *land use*.

Response to climate change: Human activities aiming at climate change adaptation and climate change mitigation.

Adaptation: Adjustment in natural or human systems to a new or changing environment. Adaptation refers to adjustments in natural or human systems, intended to reduce vulnerability to actual or anticipated cc and variability or exploit beneficial opportunities.

Mitigation: Actions resulting in reductions to the degree or intensity of GHG emissions.

Climate Change Scenario: A plausible description of how the future may develop based on a coherent and internally consistent set of assumptions about key driving forces (e.g., rate of technological change, prices) and relationships. Note that scenarios are neither predictions nor forecasts, but are useful to provide a view of the implications of developments and actions.

Sea Level Rise: The rise in the average height of the oceans over the entire globe at a single point in time. It does not include ocean tides, storm surge. Sea level rise at a specific location in the ocean may be higher or lower than the global mean because of differences in ocean temperature and other effects.

Warming of the climate system is unequivocal, as is now evident from observations of increases in global average air and ocean temperatures, widespread melting of snow and ice, and rising global average sea level.

Global warming: The rise in the average temperature of the Earth's atmosphere and oceans since the late 19th century and its projected continuation. Since the early 20th century, the Earth's mean surface temperature has increased by about 0.8 °C(1.4 °F), with about two-thirds of the increase occurring since 1980. Warming of the climate system is unequivocal, and scientists are more than 90% certain that it is primarily caused by increasing concentrations of greenhouse gases produced by human activities such as the burning of fossil fuels and deforestation. These findings are recognized by the national science academies of all major industrialized nations.

Future warming and related changes will vary from region to region around the globe. The effects of an increase in global temperature include a rise in sea levels and a change in the amount and pattern of precipitation, as well as a probable expansion of subtropical deserts. Warming is expected to be the strongest in the Arctic and would be associated with the continuing retreat of_glaciers, permafrost, and sea ice. Other likely effects of the warming include a more frequent occurrence of extreme-weather events including heat waves, droughts and heavy rainfall, ocean acidification and species extinctions due to shifting temperature regimes. Effects significant to humans include the threat to food security from decreasing crop yields and the loss of habitat from inundation.

7.2 CAUSES OF CLIMATE CHANGE

On the broadest scale, the rate at which energy is received from the sun and the rate at which it is lost to space determine the equilibrium temperature and climate of Earth. This energy is distributed around the globe by winds, ocean currents, and other mechanisms to affect the climates of different regions.

Factors that can shape climate are called "climate forcings" or "forcing mechanisms". These include processes such as variations in solar radiation, variations in the Earth's orbit, mountain-building and continental drift and changes in greenhouse gas concentrations. There are a variety of climate change feedbacks that can either amplify or diminish the initial forcing. Some parts of the climate system, such as the oceans and ice caps, respond slowly in reaction to climate forcings, while others respond more quickly.

Forcing mechanisms can be either "internal" or "external". Internal forcing mechanisms are natural processes within the climate system itself (e.g., the thermohaline circulation). External forcing mechanisms can be either natural (e.g., changes in solar output) or anthropogenic (e.g., increased emissions of greenhouse gases).

Whether the initial forcing mechanism is internal or external, the response of the climate system might be fast (e.g., a sudden cooling due to airborne volcanic ash reflecting sunlight), slow (e.g. thermal expansion of warming ocean water), or a combination (e.g., sudden loss of albedo in the arctic ocean as sea ice melts, followed by more gradual thermal expansion of the water). Therefore, the climate system can respond abruptly, but the full response to forcing mechanisms might not be fully developed for centuries or even longer.

7.2.1 Drivers of climate change

Changes in the atmospheric concentrations of GHGs and aerosols, land cover and solar radiation alter the energy balance of the climate system and are drivers of climate change. They affect the absorption, scattering and emission of radiation within the atmosphere and at the Earth's surface. The resulting positive or negative changes in energy balance due to these factors are expressed as radiative forcing, which is used to compare warming or cooling influences on global climate.

Human activities result in emissions of four long-lived GHGs: CO_2, methane (CH_4), nitrous oxide (N_2O), and halocarbons (a group of gases containing fluorine, chlorine or bromine). Atmospheric concentrations of GHGs increase when emissions are larger than removal processes.

Global atmospheric concentrations of CO_2, CH_4 and N_2O have increased markedly as a result of human activities since 1750 and now far exceed pre-industrial values determined from ice cores spanning many thousands of years (Figure 7.1). The atmospheric concentrations of CO_2 and CH_4 in 2005 exceed by far the natural range over the last 650,000 years. Global increases in CO_2 concentrations are due primarily to fossil fuel use, with land-use change providing another significant but smaller contribution. It is very likely that the observed increase in CH_4 concentration is predominantly due to agriculture and fossil fuel use. The increase in N_2O concentration is primarily due to agriculture.

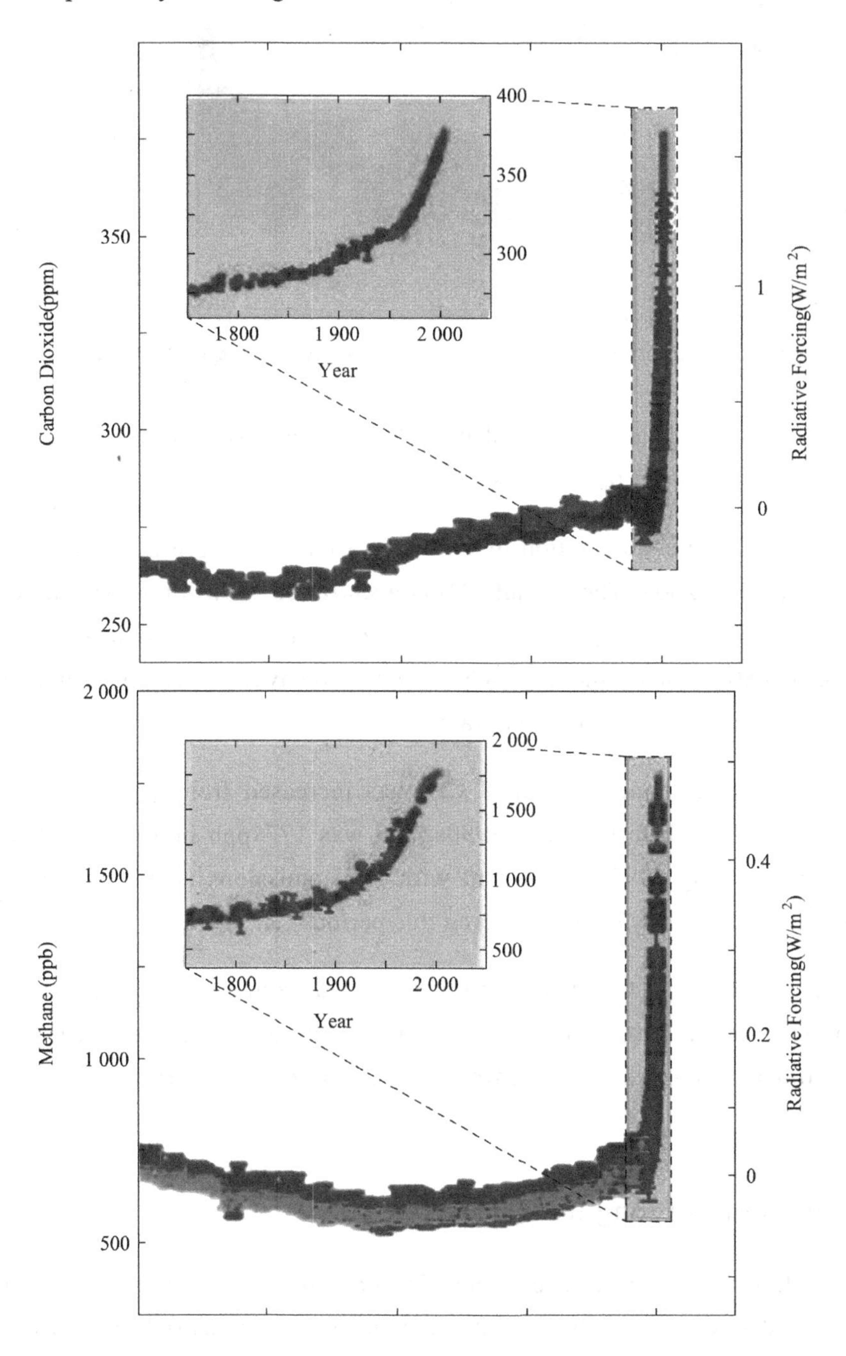

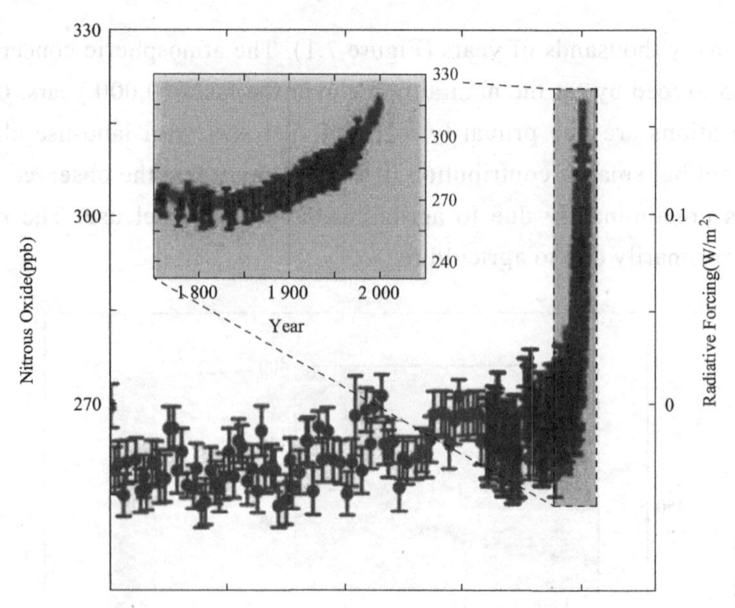

Figure 7.1 Atmospheric concentrations of CO_2, CH_4 and N_2O over the last 10,000 years (large panels) and since 1750 (inset panels)

* Measurements are shown from ice cores and atmospheric samples. The corresponding radiative forcings relative to 1750 are shown on the right hand axes of the large panels.

The global atmospheric concentration of CO_2 increased from a pre-industrial value of about 280ppm to 379ppm in 2005. The annual CO_2 concentration growth rate was larger during the last 10 years (1995-2005 average: 1.9ppm per year) than it has been since the beginning of continuous direct atmospheric measurements (1960-2005 average: 1.4ppm per year), although there is year to year variability in growth rates.

The global atmospheric concentration of CH_4 has increased from a pre-industrial value of about 715ppb to 1732ppb in the early 1990s, and was 1774ppb in 2005. Growth rates have declined since the early 1990s, consistent with total emissions (sum of anthropogenic and natural sources) being nearly constant during this period.

The global atmospheric N_2O concentration increased from apre-industrial value of about 270ppb to 319ppb in 2005. Many halocarbons (including hydrofluorocarbons) have increased from a near-zero pre-industrial background concentration, primarily due to human activities.

7.2.2 Emissions of long lived GHGs

The radiative forcing of the climate system is dominated by the long-lived GHGs. Global GHG emissions due to human activities have grown since preindustrial times, with an increase of 70% between 1970 and 2004. Carbon dioxide (CO2) is the most important anthropogenic GHG. Its annual emissions have grown between 1970 and 2004 by about 80%, from 21 to 38 gigatonnes (Gt), and represented 77% of total anthropogenic GHG emissions in 2004 (Figure 2). The rate of growth of CO_2 -eq emissions was much higher during the recent 10-year period of 1995-2004 (0.92 $GtCO_2$ -eq per year) than during the previous period of 1970-1994 (0.43 $GtCO_2$-eq per year).

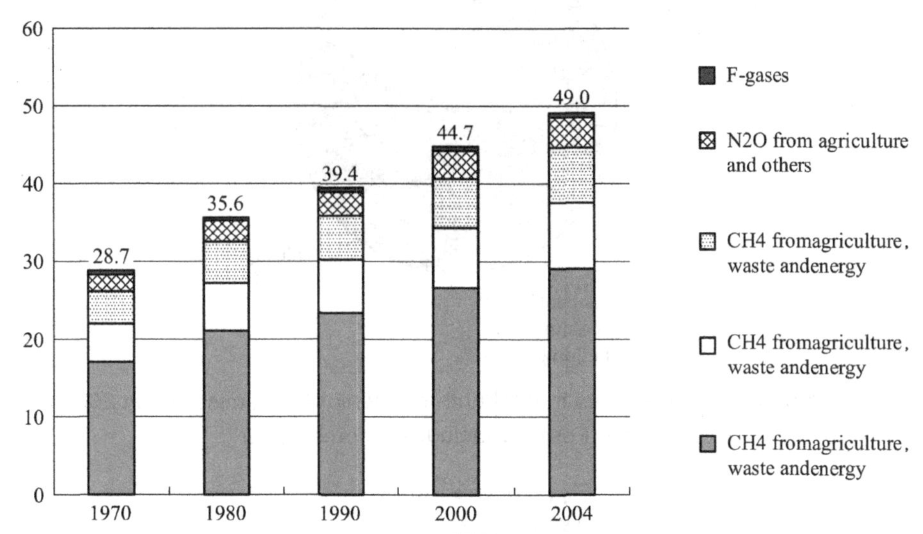

Figure 7.2 Global annual emissions of anthropogenic GHGs from 1970 to 2004

7.2.3 Carbon dioxide-equivalent (CO2-eq) emissions and concentrations

GHGs differ in their warming influence (radiative forcing) on the global climate system due to their different radiative properties and lifetimes in the atmosphere. These warming influences may be expressed through a common metric based on the radiative forcing of CO_2.

- CO_2-equivalent emission is the amount of CO_2 emission that would cause the same time-integrated radiative forcing, over a given time horizon, as an emitted amount of a long-lived GHG or a mixture of GHGs. The equivalent CO_2 emission is obtained by multiplying the emission of a GHG by its Global Warming Potential (GWP) for the given time horizon. For a mix of GHGs it is obtained by summing the equivalent CO_2 emissions of each gas. Equivalent CO_2 emission is a standard and useful metric for comparing emissions of different GHGs but does not imply the same climate change responses.

- CO_2-equivalent concentration is the concentration of CO_2 that would cause the same amount of radiative forcing as a given mixture of CO_2 and other forcing components.

The largest growth in GHG emissions between 1970 and 2004 has come from energy supply, transport and industry, while residential and commercial buildings, forestry (including deforestation) and agriculture sectors have been growing at a lower rate. The sectoral sources of GHGs in 2004 are considered in Figure 7.3.

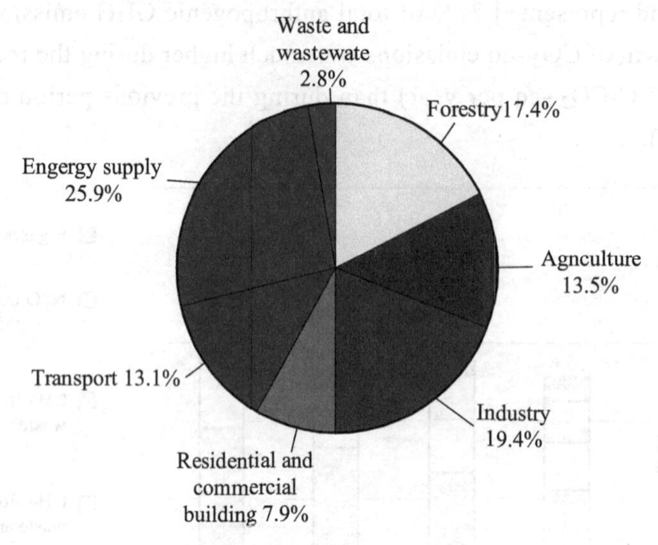

Figure 7.3 Share of different sectors in total anthropogenic GHG emissions in 2004 in terms of CO_2-eq (Forestry includes deforestation)

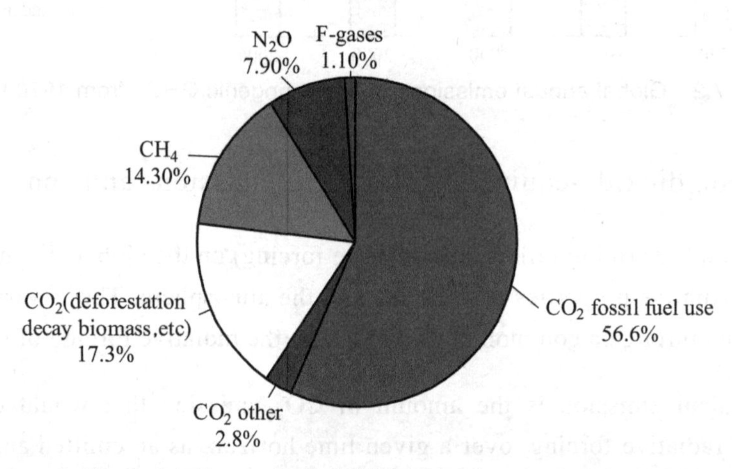

Figure 7.4 Share of different anthropogenic GHGs in total emissions in 2004 in terms of CO_2-eq

7.3 GLOBAL CLIMATE CHANGE AND VIETNAM

7.3.1 Global climate change

Warming of the climate system is unequivocal, as is now evident from observations of increases in global average air and ocean temperatures, widespread melting of snow and ice, and rising global average sea level.

Eleven of the last twelve year (1995 – 2006) ranks among the twelve warmest years in the instrumental record of global surface temperature (since 1850). The 100 year linear trend (1906–2005) of 0.74 [0.56 to 0.92]°C is large than the corresponding trend of 0.6 [0.4 to 0.8]°C (1901–2000). The temperature increase is widespread over the globe and is greater at higher northern latitudes. Average Arctic temperatures have increased at almost twice the global average rate in the past 100 years . Land regions have warmed faster than the oceans. Observations since 1961 show that average temperature of global ocean has increased to depths of at least 3,000 m and that the ocean has been taking up over 80% of the heat being added to the climate system. New analyses of ballonborne and satellite measurements of lower and mid-tropospheric temperature show warming rates similar to those observed in surface temperature.

Observational evidence from all continents and most oceans shows that many natural systems are being affected by regional climate changes, particularly temperature increases.

There is high confidence that natural systems related to snow, ice and frozen ground (including permafrost) are affected . Examples are :

– Enlargement and increased numbers of glacial lakes,

– Increasing ground instability in permafrost regions and rock avalanches in mountain regions,

– Changes in some Aratic and Antarctic ecosystems including those in sea-ice biomes, and predators at high levels of the food web.

Based on growing evidence, there is high confidence that the following effects on hydrological systems are occurring: increased runoff and earlier spring peak discharge in many glacier and snowfed rivers, and warming of lakes and rivers in many regions, with effects on thermal structure and water quality.

There is very high confidence, based on more evidence from a wider range of species , that recent warming is strongly affecting terrestrial biological systems, including such changes as earlier timing of spring events, such as leaf-unfolding, bird migration and egg-laying, and poleward and upward shifts in ranges in plant and animal species. Based on satellite observations since the early 1980s, there is high confidence that there has been a trend in many regions towards earlier "greening of vegetation" in the spring linked to longer thermal growing seasons due to recent warming.

There is high confidence, based on substantial new evidence, that observed changes in marine and freshwater biological systems are associated with rising water temperatures, as well as related changes in ice cover, salinity, oxygen levels and circulation. These include: shifts in ranges and changes in algal, plankton and fish abundance in high-latitude and high-altitude lakes , and range changes and earlier fish migrations in rivers. While there is increasing evidence of climate-related stresses from other stresses (e.g. over-fishing and pollution) is difficult.

Other effects of regional climate changes on natural and human environments are emerging, although many are difficult to discern due to adaptation and non-climate drivers.

Effects of temperature increases have been documented with medium confidence in the following managed and human systems:

Agricultural and foretry management at Northern Hemisphere higher latitudes, such as earlier spring planting of crops , and alterations in disturbances of forests due to fires and pests.

Some aspects of human health, such as excess heat related mortality in Europe, changes in infectious disease vectors in parts of Europe, and earlier onset of and increases in seasonal production of allergenic pollen in Northern Hemisphere high and mid-latitudes.

Some human activities in the Arctic (e.g hunting and shorter travel seasons over snow and ice) and in lower elevation alpine areas (such as limitations in mountain sports).

Sea level rise and human development are together contributing to losses of coastal wetlands and mangroves and increasing damage from coastal flooding in many areas.

7.3.2 Climate change in Vietnam

Viet Nam has been predicted to be one of the countries most vulnerable and likely to be significantly impacted by climate change. Globally, it has been ranked as a "Natural Disaster Hotspot", ranking 7th on economic risk, 9th on the percentage of land area and population exposed, and 22nd on mortality from multiple hazards (Dilley et al, 2005). With its "megadeltas" and high population concentrations within the Mekong and Red River Deltas, the IPCC's Fourth Assessment characterized Vietnam as a "Hotspot of key future climate impacts and vulnerabilities in Asia" (Cruz et al, 2007).

According to the World Bank (2007), Vietnam is among the countries which are hardest hit by climate change and sea level rise. Mekong and Red rivers' delta are projected to be the most seriously inundated. With sea level rise of 1 meter, about 10% of the population would be directly affected and loss of GDP would be about 10%. About 40,000km^2 of the coastal deltas are inundated every year, in which, about 90% of the Mekong River delta is almost completely inundated. If sea level rises by 3 meters, about 25% of the population would be directly affected with GDP loss of about 25%.

The annual average temperature of Vietnam has increased about 0.7 °C in the last 50 years; sea level has increased about 20cm in the same time. El-Nino and La-Nina have seriously impacted Vietnam. Climate change has made natural disasters, especially storms, floods, and droughts, become increasingly severe. It is predicted that the average temperature will increase by 3 °C and the average sea level of Vietnam will increase by 1m at the year 2010.

Climate change impacts on Vietnam are considered to be serious. It is obviously a challenge to the cause of hunger eradication and poverty reduction, the implementation of millennium development goals, and the country's sustainable development. Sectors, areas, and localities vulnerable to climate change are defined as Water resources, Agriculture and Food Security, Public Health, Deltas and coastal areas.

Moreover, according to Le Duc Nam and Le Quang Tuan (2008) and MARD (2008), extreme low temperatures in Vietnam have become more common. This means that Vietnam will also experience more severe flooding. In the next decades, these several climate extreme events are expected to become more frequent, intensive and abnormal in parallel with an expected temperature increase of 3°C by 2100.

Several studies have been done to estimate how the tropical Vietnamese climate has changed in recent decades. Nguyen Van Thang et al. (2008) used data from the past 70 years (1931–2000) to argue that the average temperature increased by 0.7 °C, which is not far from the global average. The sea-level has been rising much more compared to global averages: in a period of 50 years it rose from 25 to 30cm. Furthermore, these authors expect that Vietnam will also experience changes in precipitation rate which are different for distinct regions. For instance, the precipitation has an increase trend in Da Nang – the Central Coast, conversely a decrease trend in Ha Noi and Ho Chi Minh – the North and South, respectively. Furthermore, the authors mentioned that the frequency of storms operating on the East Sea as well as storms landing on the Vietnamese coast decreased during the last four decades from 114 storms in the 1960s to 68 storms in the 1990s. The intensity of these storms has increased and the storm season has been extended.

7.4 IMPACTS OF CLIMATE CHANGE ON ENVIRONMENT

7.4.1 Globle climate change impacts on systems

Ecosystems The resilience of many ecosystems is likely to be exceeded in this century by an unprecedented combination of climate change, associated disturbances (e.g. flooding, drought, wildfire, insects, ocean acidification), and other global change drivers (e.g. land-use change, pollution, fragmentation of natural systems, over exploitation of resources).

Over the course of this century, net carbon uptake by terrestrial ecosystems is likely to peak before mid-century and then weaken or even reverse, thus amplifying climate change.

Approximately 20 to 30% of plant and animal species assessed so far are likely to be at increased risk of extinction if increasing global average temperature exceed 1.5 to 2.5°C (medium confidence).

For increases in global average temperature exceeding 1.5 to 2.5°C and in concomitant atmospheric CO_2 concentrations, there are projected to be major changes in ecosystem structure and function, species' ecological interactions and shifts in species' geographical ranges, with predominantly negative consequences for biodiversity and ecosystem goods and services, e.g. water and food supply.

Coast Coasts are projected to be exposed to increasing risks, including coastal erosion, due to climate change and sea level rise. The effect will be exacerbated by increasing human-induced pressures on coastal areas (very high confidence).

By the 2080s, many millions more people than today are projected to experience floods every year due to sea level rise. The numbers affected will be largest in the densely populated and low-lying megadeltas of Asia and Africa while small islands are especially vulnerable (very high confidence).

Water Water impacts are key for all sectors and regions. Climate change is expected to exacerbate current stresses on water resources from population growth and economic and land-use hange, including urbanisation. On a regional scale, mountain snow pack, glaciers and small ice caps play a crucial role in freshwater vailability. Widespread mass losses from glaciers and reductions in snow cover over recent decades are projected to accelerate through out the 21st century, reducing water availability, hydropower potential, and changing seasonality of flows in regions supplied by meltwater from major mountain ranges (e.g. Hindu-Kush, Himalaya, Andes), where more than one-sixth of the world population currently lives.

Changes in precipitation and temperature lead to changes in runoff and water availability. Runoff is projected with high confidence to increase by 10 to 40% by mid-century at higher latitudes and in some wet tropical areas, including populous areas in East and South-East Asia, and decrease by 10 to 30% over some dry regions at mid-latitudes and dry tropics, due to decreases in rainfall and higher rates of evapotranspiration. There is also high confidence that many semi-arid areas (e.g. the Mediterranean Basin, western United States, southern Africa and north-eastern Brazil) will suffer a decrease in water resources due to climate change. Drought-affected areas are projected to increase in extent, with the potential for adverse impacts on multiple sectors, e.g. agriculture, water supply, energy production, and health.

Regionally, large increases in irrigation water demand as a result of climate changes are projected.

The negative impacts of climate change on freshwater systems outweigh its benefits (high confidence). Areas in which runoff is projected to decline face a reduction in the value of the services provided by water resources (very high confidence). The beneficial impacts of increased annual runoff in some areas are likely to be tempered by negative effects of increased precipitation variability and seasonal runoff shifts on water supply, water quality, and flood risk.

7.4.2 Impacts of Climate Change in Vietnam

Vietnam has a long coastline of about 3,260km, million square kilometres of water shelf, and more than 3,000 off-shore islands, large coastal low land areas. Therefore, Vietnam experiences high floods in rainy season and droughts and salt intrusion in dry season. Climate change and sea level rise would make these risks more serious, increase flooded areas, obstruct water drainage, intensify coastal line erosion and salt intrusion which causes difficulties for agricultural production and domestic water usage, and create critical risks to coastal infrastructures such as sea malnutrition, road, docks, and factories, urban areas and coastal communities. Sea level and sea water temperature rising have potential adverse effects on coral reefs and mangrove forests, biological foundations which are bases for coastal aquaculture and fishery. Therefore, significant investments should be attracted into sea-dyke construction and consolidation to respond to sea level rise, infrastructure development, resettlement of coastal communities, and construction of urban areas which have high adaptability to sea level rise.

Increasing temperature has potential impacts on natural ecosystems. It would cause shifts in thermo-border of continental ecosystems and fresh water ecosystems as well as shifts in flora and fauna structure in certain regions. Degradation of biodiversity would accelerate due to loss of some temperate and sub-tropical species.

The increase of climatic extremes in both frequency and intensity due to climate change is a frequent risk, both short-term and long-term, to all sectors, regions, and communities. Storms, floods, droughts, heavy rains, and high temperature are annual disaster in many parts of the country, causing large damages to production and life. Climate change would make those natural disasters much more severe, even become catastrophes, posing risks to socio-economic development and clear up achievements of many years of development, including achievements of millennium development goals. Regions/areas which are expected to suffer biggest impacts of those extremes are coastal zone along the Central Part, mountainous region in the North and Northern Center, Red River Delta and Mekong River Delta.

Water Resources Annual flows of rivers in the North and northern areas of the North Central Coast are expected to increase. In contrast, annual flows of rivers in the southern area of North Central Coast to the northern area of South Central Coast are expected to decrease. Flood peaks in most rivers will increase while dry season low flows will decline. Potential evapotranspiration will show rapid increases in the South Central Coast, with the Mekong Delta regions having the greatest increases. After 2020, groundwater levels are expected to drop drastically. Increases in the incidence of severe drought, especially during the dry season, and in inter- and intra-seasonal rainfall patterns will create much greater uncertainty as to moisture availability, inter-annually for crop and livestock production and intra-annually for overall availability of water resources. Increases in the incidence of extreme rainfall events will exacerbate the impacts of flooding (inundation and flash floods) as well as concentrate rainfall within shorter time periods leading to decreases in soil and groundwater recharge, particularly in uplands and sloping areas and potentially greater crop moisture stress.

Agriculture Total annual temperature is projected to increase between 8% and 11% by 2100. In

most regions, the number of days when temperatures exceed 25 °C will increase notably while the number of days when temperatures drop below 20 °C will decrease. Water demand for agriculture may increase two or three-fold compared with that of 2000. Tropical plants will tend to shift further north and towards higher altitudes. Crop water shortage would be exacerbated with decreased coverage of hygrophytes and rising evapotranspiration rates. Spring crop outputs are set to decline at a faster rate than summer crop outputs. Winter maize productivity may increase in the Red River Delta but decrease in Central Coast and the Mekong River Delta. Climate change may also threaten the life cycle (i.e. growth and reproduction) of cattle and increase the incidence and spread of diseases.

Forestry Climate change will have a diverse range of impacts on forest ecosystems and flora. By 2100, native forest cover comprised of closed tropical moist semi-deciduous forests and closed evergreen forests, amongst others, will decrease. The ecosystems of closed tropical moist semi-deciduous forests are likely to be most affected by climate change.

In 2100, Chukrasia tabularis forests are projected to cover only 0.3 million ha, a decrease of 70% decrease. Pinus merkusii forests, are projected to cover approximately 2.3 million ha, equivalent to a fall of 58%. Climate change will heighten risks of forest fires in all regions, primarily during the dry-hot season. In addition, warmer conditions will facilitate the spread of forest pests, hampering the growth of forest ecosystems.

Aquaculture Climate change adversely impacts the ecosystems of coral reefs, maritime, and estuarine sea grass beds, and causes reductions in fish stocks. Sea-level rise would exacerbate salinization in coastal zones, causing the retreat of mangrove forests with accompanying losses in habitat for numerous species.

Furthermore, the advance of saltwater leads to the replacement of freshwater species by their brackish and saline water counterparts in estuaries and coastal lagoons. Finally, rising temperatures weaken aquatic species and foster the growth of harmful microorganisms.

Energy and transportation Rising temperatures will raise energy consumption for climate-sensitive sectors. Electricity transmission and distribution networks, along with oil-rigs, oil and LNG Pipe lines and shipments will be negatively impacted by rising sea levels and extreme weather. Hydroelectric power generation will be affected by changing river flows, posing new challenges to the management of reservoirs.

Due to rising sea levels, industrial facilities, equipments, power stations, and transmission lines in coastal zones face the risk of flooding. This would increase maintenance and repair costs, and affect energy supply, consumption, and national energy security. A rise of 100 cm in sea level could lead to the submersion of 11,000 km of road infrastructure, paralyze the country's transportation activities and cause considerable damage to the economy.

Human health Climate change impacts human health directly through changing climate conditions, abnormal heat waves, and increased occurrence of natural disasters. Indirectly, rising sea level and temperature affect agricultural land, food security, and increase the risks of food shortages while warmer conditions facilitate the spread of infectious diseases and nepidemics.

7.5 RESPONSE TO CLIMATE CHANGE IN VIETNAM

Viet Nam recognized the threat posed by human-induced climate change by ratifying the UNFCCC in 1994 and the Kyoto Protocol in 2002. To date, the Government has mainly focused on inventories and the reduction of green house gas emissions. The Initial National Communication (INC) to the UNFCCC (MoNRE 2003) only explored climate change impacts and necessary adaptation measures in a preliminary and qualitative way. A series of sector assessments were made and adaptation options identified, but these did not include socioeconomic analysis, and they have not yet been followed by specific programmes. More indepth vulnerability and adaptation assessments and the preparation of a policy framework for implementing adaptation measures are currently being undertaken for the Second National Communication (SNC) to the UNFCCC, which should be completed by 2009.

The Ministry of Natural Resources and Environment (MoNRE) is the national focal agency for climate change related activities, and an organogramme of MoNRE's climate change related offices is presented. Climate change adaptation measures have been included in a number of recent laws and strategies, such as the National Strategy for Environmental

Protection (2005), which includes measures for reducing the impact from sea level rise in coastal zones. In early 2006, the MoNRE-based International Support Group on Natural Resources and Environment (ISGE) established a climate change adaptation working group, which provides a forum for dialogue and should promote coordination for climate change adaptation measures.

Viet Nam already has an extensive long-standing institutional response system for natural disasters such as floods and typhoons, reflecting the country's vulnerability to these events. Disaster risk management activities are coordinated primarily by the Central Committee for Flood and Storm Control (CCFSC, founded in 1955), chaired by the Minister of Agriculture and Rural Development. Other members of the CCFSC include relevant line ministries, the Department of Floods and Storm Control and Dyke Management, the Disaster Management Centre, the Hydro-meteorological Service, and the Viet Nam Red Cross (VNRC). The Natural Disaster Mitigation Partnership (NDM-P) is made up of Government, NGOs and donors to promote dialogue and common ways of working, and support coordination for implementation of the Second National Strategy and Action Plan for Disaster Mitigation and Management.

The CCFSC is responsible for gathering data, monitoring flood and storm events, issuing official warnings, and coordinating disaster response and mitigation measures. The authorities in all localities and each sector ministry also have committees for flood and storm control (CFSCs). Local CFSCs at the provincial, district and commune levels are responsible for coordination of flood and storm measures; organising dyke protection, flood and storm preparedness and mitigation; and flood recovery and rehabilitation (EU/MWH 2006). Sector committees support with technical assistance, materials and equipment. The system of CFSCs is important for sharing information on damage and also relief needs, communicating early warning information, damage assessments, co-ordinating rescue during floods, and protecting dykes and other infrastructure. Viet Nam's mass organizations are also crucial in disaster response, with the Fatherland Front raising and dispersing considerable relief funds and supplies, for example during the 2000 and 2001 floods in the Mekong Delta (IFRC 2002). The VNRC is operating throughout the country from national to commune level and works on awareness raising, disaster preparedness, response, and prevention.

Viet Nam's policy framework for disaster management is set in the Second National Strategy and Action Plan for Disaster Mitigation and Management 2001-2020. This Strategy prioritises increased awareness raising and participation, minimizing loss of life and assets, and streses the importance of co-existence with floods in situations which demand it. Other key initiatives of the Second National Strategy include: establishment of disaster forecast centres in the north, centre and south of the country (for different disasters); construction of flood corridors and flood retention areas in southern Viet Nam; the use of advanced information and communication technology; strengthening the role of schools and the media in awareness raising; maintaining and upgrading equipment for local Flood and Storm Control Committees; and a proposal for a national disaster fund for projects on disaster mitigation and preparedness, and setting up a disaster insurance company.

The Second National Strategy is still, however, designed principally to address short-term climate extremes rather than to respond to future climate change, and focuses on emergency response and reconstruction, rather than risk prevention and adaptation. There is also a marked lack of integration between disaster risk reduction policies and wider policies for rural development and poverty reduction, with little cross sectoral integration or coordination, either in policy, or in practice. A recent study into institutional arrangements for climate change response concludes that, "Integration of institutions engaged in disaster management, climate risk and development remains a weakness in Viet Nam, but there are positive examples of coordination to build upon, including the multi-scale framework provided by the CFSC system and the NDM partnership for Central Viet Nam" (EU/MWH 2006, p27). There is limited Government ownership yet of an adaptive approach to future climate related risks, and limited financing available for climate change adaptation.

In addition, the National Target Program to Respond to Climate Change (NTP-RCC) is the umbrella program and guiding framework for the Government of Viet Nam's efforts in adaptation and mitigation of climate change risk. The Ministry of Natural Resources and Environment developed the program and is responsible for its implementation. The current program, which covers the period from 2009 to 2015, has the global objectives of: (i) assessing potential impacts of climate change; (ii) ensuring that a climate change response action plan is developed by each sector; (iii) initiating efforts to move the country towards a low-carbon economy, and (iv) contributing to global efforts for the mitigation of GHGs.

The NTP-RCC primarily represents only a first step in what will be a much longer process for taking the broad vision that it provides for the entire country and the affected sectors and translating it into specific priorities, strategies and action plans. The agenda it puts forward is primarily one of research, planning, communication, and inter-institutional and inter-sectoral coordination efforts. The development of detailed and explicit climate change responses are left, in the first instance, for the other ministries to specify through their own sectoral action plans. The NTP simply establishes the requirement that all other line ministries, provinces and cities produce climate change action plans. As such, the NTP-RCC is not a climate change strategy, though reportedly MoNRE is now working on the preparation of such a strategy.

Education, training, and public awareness-raising

Major national educational and training institutions have begun incorporating climate change studies in their official curricula for specialized students.

Awareness raising activities have been gradually broadened in both content scope and participant diversity. Several major climate change-related publications (books, periodicals, leaflets, video clips) have been translated, published and widely distributed. A climate change journal is published periodically. A number of specialized websites covering climate change issues, such as www.noccop.org.vn, www.vacne.org.vn and www.nea.gov.vn have been set up to provide timely global and national news updates.

Climate change information has been disseminated via the mass-media nationally and locally. In addition, climate change issues have been regularly mainstreamed into news topics, social events and captured much interest throughout Vietnamese society, particularly the younger generations. The Viet Nam NGO Climate Change Working Group was established in February 2008 to facilitate inter-agency coordination and foster discussions between NGOs on climate change.

Many NGOs have shown interest in climate change response activities as reflected by the organization of numerous workshops, training courses, educational games, and competitions to different target groups.

Enhancement of International Cooperation

Enhancement of international cooperation aims at: 1) taking opportunity to obtain and effectively use supports from international community, including financial support and technological transfer, through bilateral and multilateral cooperation channels; and 2) Participating in regional and global cooperation activities on climate change. Targets to be achieved are: a cooperation mechanism between Viet Nam and international donors in implementing the NTP established; Bilateral and multilateral cooperation between Viet Nam and some other countries/international organizations to respond to climate change established; initial aid (financial support, technology and expertise transfer) of international community to Viet Nam to implement the NTP primarily identified; Vietnam to contribute to the development of international agreements/documents on climate change after 2012; a framework of legal documents on mechanisms/policies to encourage investment into CDM projects, climate change response projects, and environmental friendly technology transfer projects supplemented and finalized so as to facilitate foreigner partners to invest into such projects in Vietnam.

Specific Viet Nam's Current Climate Change Responses & Adaptation

Coastal Defences: Dyke Management and Mangrove Restoration

Physical protection from typhoons and rising sea levels is provided by Viet Nam's extensive system of dykes – 5,000 km of river dykes and 3,000 km of sea dykes. Dykes and levees have existed for over 1000 years. Local government remains responsible for sea dyke protection. In the past there was an extensive system of labour contributions for building and maintaining dykes, but this has increasingly been replaced by a system of hired labour and local taxes. An Oxfam GB programme in Ky Anh district, Ha Tinh province in the 1990s showed that support provided to local communities in organising and mobilising for sea dyke strengthening and maintenance improved collective security and enabled local people to invest in improving the productivity of their land. This provided a viable alternative to outmigration for vulnerable coastal communities. Coastal mangrove plantation is also an important and highly effective form of coastal protection from storm surges following tropical storms and cyclones. As an illustration, it is estimated that in Kien Thuy District, a 4-metre high storm surge from storm number 7 in 2005 (typhoon "Damrey" – see annex) was reduced to a 0.5m wave by extensive restored mangrove (Jegillos et al. 2005 in EU/MWH 2006). Both international donors and NGOs have successfully supported coastal communities in mangrove restoration.

Disaster Early Warning Systems

Disaster warning and preparedness is a key aspect of Viet Nam's response to climate related threats and disasters. UNDP has long supported Viet Nam in improving early warning for disasters, gathering and reporting damage data, and in connecting Viet Nam's hydrometeorological data services and CCFSC to the national media in order to make information more readily and more widely available. The Government is continually upgrading capacity in this regard and satellite data are expected to be available in 2008 from Viet Nam's own satellite, Vinasat. Real time meteorological information is also available from China and Japan's Meteorological agencies, but improvements in information collection and communication are especially needed to prevent the large loss of life through sinking of boats as occurred in the East Sea in 2006 during typhoon Chanchu (see annex). The national typhoon warning system delivers a 48-hour warning, broadcast through the media and locally via loudspeakers, and during the typhoon season dykes are monitored 24 hours a day (EU/MWH 2006). The CCFSC also disseminates reports by electronic mail. Nevertheless, despite recent improvements the system is still in need of improvement.

Appendix

Special Words and Phrases

A

abate
v. 减轻，减弱

abiotic
a. 无生命的，非生物的

abrade
v. 磨损

abscissa
n. 横坐标

absolute humidity
绝对湿度

absorption
n. 吸收，吸收作用

abundant
a. 大量的，丰富的

abuse
n. 滥用
v. 滥用

accelerate
v. 加速，促进

accomplish
v. 完成，实现

accretion
n. 添加，添加物，停滞堆积

accuracy
n. 精度，精确，准确度

acid
n. 酸，酸类物质
a. 酸的，酸性的

acidic
a. 酸的，酸性的

acidification
n. 酸化，使发酸

acoustical
a. 听觉的，声学的

acute
a. 严重的，急性的，尖锐的

adequate
a. 适当的，足够的

adhesive
a. 粘着的
n. 粘合剂

adiabat
n. 绝热线

adiabatic
a. 绝热的

adsorption
n. 吸附，吸附作用

aerate

v. 曝气，充气

aerobic

a. 需氧的

aerodynamic

a. 空气动力学的

aerosol

n. 烟，雾，气溶胶

agency

n. 中介，代理机构

agent

n. 剂，媒介

aggravate

v. 使恶化，加重

aggressiveness

n. 侵犯，过分

agitate

v. 搅动，搅拌

agrarian

a. 土地的，耕地的

airborne

a. 空气传播的，空运的

alcohol

n. 酒精，乙醇

alfalfa

n. 苜蓿

algae

n. (pl.) 藻类

algal

a. 海藻的

alien

n. 外国人

a. 相异的，异己的

alignment

n. 校正，对中，列队，结盟

alkaline

a. 碱的，碱性的

alkyl

n. 烷基，烃基

a. 烷基的，烃基的

allocate

v. 分配

aloft

ad. 在上，在高处

alveoli

n. 气泡，齿槽突起

ambient

a. 周围的

n. 周围环境

ameliorate

v. 改善，改进

amenable

a. 服从的，有义务的

amino

a. 氨基的

ammonia

n. 氨，氨水

ammonification

n. 氨化，氨化作用

ammonium

n. 铵

amphibian

a. 两栖类的，水陆两用的

n. 两栖动物，水陆两用飞机

ample

a. 充足的，丰富的，强大的

anaerobic

a. 厌氧的

anaerobic digester

厌氧消化池

analogous

a. 类似的，相似的

ancestral

a. 祖传的，祖先的

anoxic

a. 缺氧的

antacid

a. 抗酸的

n. 抗酸剂

anthropoid

a. 似人的

n. 类人猿

anthropogenic

a. 人类发生的，人类起源的

anthropology

n. 人类学

anticipate

v. 预先做出，预见，提前进行

anticyclone

n. 反气旋，高气压

antique

n. 古物

a. 旧式的，过时的，古老的

anvil

n. 铁砧

aquatic

a. 水生的，水上的

n. 水生动物，水草

aqueduct

n. 水管，沟渠

aqueous

a. 水的，似水的

aquifer

n. 含水层，蓄水层

arable

a. 可耕的

archaeological

a. 考古学的

arduous

a. 艰苦的，费力的

arid

a. 干旱的，贫瘠的

arithmetic

n. 算术

array

n. 排列，大批

v. 排列，布署

asbestos

n. 石棉

ascend

v. 上升，攀登

aspiration

n, 渴望，热望

assault

n. 攻击，袭击

v. 攻击，袭击，发动攻击

assimilate

v. 吸收

assimilation

n. 同化，同化作用，吸收

assimilative

a. 同化的

assortment

n. 分类

astronomical

*a.*天文学的，庞大无法估计的

n. 天文

atmospheric

a. 大气的

attenuate

a. 稀薄的，减弱的

v. 使变弱，减少

attenuation

n. 衰减，稀薄化

attest

v. 证明，证实

auditor

n. 旁听者，审计员

auditory

a. 听觉的，耳的

n. 礼堂

autotroph

n. 自营生物

auxiliary

n. 辅助品，附属设备

a. 辅助的，备用的

avid

a. 渴望的，急切的

awkward

a. 笨拙的，棘手的，难处理的

awry

a. 错误的

ad. 歪斜的

axial

a. 轴的，轴向的

axis

n. 轴

azotobacter

n. 固氮菌类

B

bacillus

n. 杆菌，细菌

backlash

n. 反撞，齿轮隙，后冲

bacteriological

a. 细菌学的

bacteriophage

n. 抗菌素

baffle

v. 为难，困惑

n. 挡板，折流板

bale

n. 包，捆，灾难，货物

v. 打包，捆包

balloonist

n. 驾驶/操纵气球的人

ban

n. 禁令

v. 禁止，取缔

bar rack

n. 条棒格栅

batch

n. 一次操作所需的原料量

v. 分批，分批制作

a. 分批的，间歇式的

beehive

n. 蜂窝，蜂箱

beet

n. 甜菜

v. 修理，改过

benzene

n. 苯

biochemical

a. 生物化学的

biodegradable

a. 可生物降解的

biogeochemical cycle

n. 生物化学循环

magnification

n. 放大率，扩大

biomass

n. 生物量，生物质

biome

n. 生物群落区

biosphere

n. 生物圈

biota

n. 生物区

biotic

a. 生物的，生命的

birch

n. 桦树，桦木

blend

v. 混合

n. 混合

block

n. 块，石块，木块

v. 阻塞，妨碍

boom

n. 繁荣，隆隆声

v. 兴旺，迅速发展

boreal

a. 北的

borough

n. 自治的市镇，区

botanical

a. 植物学的

n. 植物性药材

bounty

n. 慷慨，宽大，奖金

bowl

n. 碗，槽，球形物

v. 滚动

brittle

a. 易碎的，脆弱的，易坏的

broaden

v. 变宽，扩大，放宽

broadness

n. 宽度，明白

brook

n. 小溪

v. 容忍

brownish

a. 呈褐色的

bubble

n. 泡沫

v. 冒泡，沸腾

buffer

n. 缓冲器，缓冲区

v. 缓冲

buildup

n. 建造，集结，组成，增强，形成

bulk

n. 大小，体积，大块，大多数

v. 显得大，显得重要

bulky

a. 庞大的

buoyancy

n. 浮力，弹性

bureaucracy

n. 官僚，官吏

burgeon

n. 嫩芽

v. 萌芽

C

cabinet

n. 橱柜，内阁

a. 内阁的

cactus

n. 仙人掌

cadmium

n. 镉

calcium

n. 钙

cam

n. 凸轮

canister

n. 罐，筒，滤毒罐

cannery

n. 罐头厂

canyon

n. 峡谷，深谷

carbohydrate

n. 碳水化合物，糖类

carbon dioxide

n. 二氧化碳

carbon monoxide

n. 一氧化碳

carbonaceous

a. 含碳的，碳质的，碳的

carbonate

a. 碳酸盐 (酯)

v. 使碳酸盐化，使充满二氧化碳

carburetor

n. 汽化器，增碳器

carcass

n. 尸体

carcinogen

n. 致癌物

carcinogenic

a. 致癌的

cardboard

n. 薄纸板，特等薄纸，卡片纸板

cardiac

n. 强心剂

a. 心脏的

cardiovascular

a. 心血管的

carnivore

n. 食肉动物

carp

n. 鲤鱼

v. 挑剔

carrion

n. 死肉，腐肉

a. 吃腐肉的

catalyst

n. 催化剂

catalytic

a. 催化的

catastrophe

*n.*大灾难，大祸

cattail

n. 柔荑花，香蒲

caulk

v. 填……以防漏

caveman

n. 穴居人

cavity

n. 洞，空穴，腔

cedar

n. 雪松，香柏，杉木

cellular

a. 细胞的

cellulose

n. 纤维素

centrifugal

a. 离心的

n. 离心机

centrifugation

n. 离心法，离心过滤

centrifuge

n. 离心机

v. 使受离心作用

certification

n. 证明

chamber

n. 室，房间

v. 装入室中

a. 室内的

chaos

n. 大混乱，混沌

charcoal

n. 木炭

cherish

v. 爱护

chestnut

n. 栗子，栗树，栗色

a. 栗色的

chip

n. 屑片，薄片，碎片

v. 削，切，削成碎片

chipmunk

n. 花栗鼠

cholera

n. 霍乱

chore

n. 零工，家务

chromate

n. 铬酸盐

chromium

n. 铬

chronic

a. 慢性的，长期的

chug

n. 嘎嚓声

v. 发出嘎嚓声

chute

n. 斜槽，瀑布

clam

n. 蛤

clapper

n. 铃锤，警钟锤

clarifier

n. 澄清池，净化剂

clearance

n. 清除，间隙，空隙

climatology

n. 气候学

cling

v. 粘紧，附着，坚持

clostridium

n. 梭状芽胞杆菌

coarse

a. 粗糙的

cob

n. 圆块

v. 捣碎

cochlea

n. 耳蜗

coefficient

n. 系数

coincide

v. 一致，符合

coliform

n. 大肠菌

a. 大肠杆菌状的

coliphage

n. 大肠菌噬体

collision

n. 碰撞，冲突

colloidal

a. 胶态的，胶体的，胶质的

colonization

n. 殖民，殖民地化

colony

n. 菌落，殖民地，群体

colossal

a. 巨大的

combustion

n. 燃烧

commensalism

n. 共生，共栖，共生现象

comminutor

n. 粉碎机

community

n. 群落，团体，社会，地区

comparative

a. 相当的，比较的

n. 对手

complaint

n. 诉苦，抱怨

composite

a. 复合的，合成的

n. 合成物
compost
n. 堆肥，混合肥料
v. 堆肥于
compromise
n. 折中，和解
v. 妥协，折中
condensation
n. 冷凝，浓缩，凝聚
conduit
n. 水管，沟渠，导管
configuration
n. 结构，构造，配置
coning
n. 圆锥角，形成圆锥形
a. 圆锥形的了
conjunction
n. 联合，结合，连接词
conscientious
a. 有责任心的，负责的
consolidate
v. 巩固，联合，统一
consumer
n. 消费者
contactor
n. 接触器
contaminate
v. 污染，毒害
contamination
n. 污染，污物
continent
n. 大陆，洲
contour
n. 轮廓，外形，等高线
v. 画轮廓
a. 使与轮廓相符的
controversy
n. 争论，辩论，争吵
convalescence
n. 恢复期
conversion
n. 转化，转换
convey
v. 运输，传达
conveyor
n. 运输机，传送带，输送设备
copper
n. 铜，警察
coriolis acceleration
互补 (复合向心) 加速 (度)
cornerstone
n. 奠基石，基础
corona
n. 电晕，冠，光圈
corrosion
n. 腐蚀，锈蚀，侵蚀
corrugate
v. 使起皱，起皱，成波状
a. 有皱的，波纹状的
coyote
n. 丛林狼，讨厌的家伙
crack
n. 裂缝，龟裂，爆裂声
v. (使) 爆裂, (使) 裂开
a. 最好的
creep
n. 蠕变，爬
v. 爬，蠕变，徐行
crest
n. 冠，饰毛
v. 到达绝顶，加以顶饰
cricket
n. 蟋蟀，板球
criteria
n. (pl.) 标准
crucible

n. 坩锅，严酷的考验

crustacean

a. 甲壳类的

n. 甲壳类动物

cube

n. 立方体，立方

cumulative

a. 累积的，蓄积的

curb

n. 路边，控制，抑制

v. 抑制，束缚

curtail

v. 缩减，剥夺，简略

curvature

n. 弯曲，曲率

cushion

n. 垫子，弹性垫，软弹性垫

v. 缓冲

cutoff

n. 切断，截止，捷径

cyanate

n. 氰酸盐

cyanide

n. 氰化物

cyclic

a. 周期的，循环的，环状的

cyclone

n. 离心式除尘器，龙卷风

cylinder

n. 圆柱 (体)，量筒，气缸

cylindrical

a. 圆柱体的，圆筒形的

cyst

n. 囊肿，包囊

D

dampen

v. 弄湿，使沮丧，丧气，使减幅

damper

n. 减震器，阻尼器

dearth

n. 缺乏，供应不足，饥荒

debilitate

v. 使衰弱

debris

n. 碎片，残骸，碎屑

decay

n. 衰退，腐蚀，腐烂

v. (使) 腐败，(使) 衰退

decay organism

腐烂有机体

decelerate

v. 减速，放慢

decibel

n. 分贝

decimation

n. 大批杀害

decomposer

n. 分解者

dedication

n. 奉献，捐献

deformity

n. 畸形，残缺，畸形的人或物

defrost

v. 除霜，解冻

demise

*n.*死亡，禅位

v. 遗赠，转让

demister

n. 除雾器，去雾器

denitrification

n. 反硝化作用，脱氮作用

denitrifying bacteria

脱氮细菌

dense

a. 密集的，浓厚的

depict
v. 描述，描写
depletion
n. 消耗，耗尽
deposition
n. 沉积作用，沉积物，免职
descriptive
a. 描述的，叙述的
destiny
n. 命运，定数
detention
n. 停滞，阻止，滞留
detergent
n. 清洁剂，去垢剂
deterioration
n. 恶化，衰退，退化，变质
detritus
n. 碎石，腐质
devastate
v. 毁坏，使荒废，掠夺
deviation
n. 偏差，误差，背离
devise
v. 设计，发明，遗赠给
n. 遗赠
dewater
v. 使脱水
diarrhea
n. 腹泻
diatom
n. 硅藻属
dichromate
n. 重铬酸盐
diffract
v. 使分散，衍射
digestive
a. 消化的，助消化的
n. 助消化药
dilate
v. 扩大，使膨胀
dilute
v. 冲淡，稀释
a. 淡的，稀释的
dilution
n. 稀释，稀释法
dimension
n. 尺寸，维数，量纲
v. 标出尺寸
diminution
n. 减少，缩小，减小
dine
v. 进餐，宴请
dinosaur
n. 恐龙
dire
a. 可怕的，悲惨的，极端的
disc
n. 圆盘，轮盘，唱片
v. 灌唱片
discard
v. 抛弃
n. 废品
discharge
v. 排出，流出，卸货
n. 排放量，排出物
disinfection
n. 消毒
dislodge
v. 逐出，使移动，离开原位
dispersion
n. 分散，分散作用，散布
disposal
n. 处理，处置
dispute
n. 争论，辩论，争吵
v. 争论，辩论，怀疑

dissect

v. 解剖，切开

dissipate

v. 消散，驱散，浪费

dissociate

v. 分离，游离，分裂

dissolved oxygen

溶解氧

dissolved solids

溶解固体

distort

v. 扭曲，歪曲

diurnal

a. 每日的，白天的

n. 日报，日刊

divergence

n. 发散，分歧

diversity

n. 多样性，差异

division

n. 分开，除法

docile

a. 容易教的，温顺的

document

n. 文件，文档，公文

v. 证明，为……引证

domain

n. 领域，范围

domestic

a. 家庭的，国内的

doom

n. 厄运，死亡

*v.*注定要

dosage

n. 剂量，用量

dose

n. 剂量

v. （给……）服药

downspout

n. 溢流管

draft

n. 气流，草稿，计划，取水量

v. 起草，征集

drainage

n. 排水，排水装置

drawback

n. 缺点，障碍，退税

dredge

n. 挖泥机，拖网，砂耙

v. 挖掘，疏浚

drizzle

n. 细雨

v. 下毛毛雨

droplet

n. 小滴

drought

n. 干旱，缺乏

drum

n. 转筒，卷盘，卷筒，鼓

v. 击鼓

duct

n. 管，导管，输送管

v. 通过管道输送

durable

a. 持久的，耐用的

dwindle

v. 减少，使缩小

dynamic

a. 动态的，动力的

n. 动态，动力

dynamic equilibrium

动态平衡

dysentery

n. 痢疾

E

eardrum

n. 中耳，鼓膜，鼓室

ecological

a. 生态的，生态学的

ecological balance

生态平衡

ecological pyramid

生态金字塔

ecology

n. 生态学

ecosystem

n. 生态系统

eddy

n. 旋涡

v. (使) 起旋涡

effluent

a. 流出的，射出的

n. 排放物，污水

eggshell

n. 蛋壳

egret

n. 白鹭

elaboration

n. 精巧，详细描述，苦心经营

electrode

n. 电极，电焊条，焊条

elevation

n. 提高，海拔，高程，正视图

elimination

n. 消除，排除，除去，消灭

elk

n. 麋鹿

elliptical

a. 椭圆的

ember

n. 灰烬，余烬

embryo

n. 胚胎，胎儿，胚芽

a. 胚胎的，初期的

emerald

n. 翡翠，翠绿色，绿宝石

a. 翡翠的，翠绿色的

emission

n. 发射，发行

enclosure

n. 外壳，箱，围墙，附件

endogenous

a. 内源的，内生的

entourage

*n.*随行人员，周围，环境

entrain

v. （使）坐火车，输送，导致

entrainment

n. 输送，夺取，雾沫，挟带

environment

n. 环境

environmental disturbance

环境干扰，环境失调

environmental phase

环境相

environmental resistance

环境耐力，环境容量

epidemic

n. 传染病，流行病

a. 传染的，流行的

episode

n. 插曲，一段情节

equalization

n. 平衡，均衡，均等

equator

n. 赤道

equilibrium

n. 平衡

eradicate

v. 根除，消灭，根绝

erosion

n. 腐蚀，侵蚀，磨耗

estuarine

a. 河口的，港湾的

etch

v. 蚀刻，蚀镂

n. 腐蚀剂

ethical

a. 伦理的，道德的

eutrophication

n. 富营养化

evaporation

n. 蒸发 (作用)，脱水

evapotranspiration

*n.*土壤水分蒸发蒸腾损失总量

evidence

n. 根据，证据

excess

n. 过度，过量，超额

a. 过量的，额外的

excretion

n. 排泄，排泄物

exhibit

n. 显示，陈列品，证据，证件

v. 表现， 展览

expansion

n. 扩充，扩建，膨胀

expedient

n. 权宜之计

a. 权宜的，方便的，有用的

exploitive

a. 开发的，利用的，剥削的

exponential

a. 指数的

n. 指数

extraction

n. 萃取法，提取，抽出

extractive

a. 萃取的，抽取的

n. 抽出物

extravagant

a. 奢侈的，浪费的

F

fabricate

v. 制造，建造，装配，伪造

facilitate

v. 使容易，促进，帮助

facility

n. 容易，设备，装备，工具

facultative

a. 兼性的，可选择的

faint

n. 晕厥，昏厥，昏倒

a. 模糊的，无力的

v. 昏倒，变得微弱

falcon

n. 隼，猎鹰

fatality

n. 不幸，厄运，致命性

fecal

a. 排泄物的，粪便的

feces

n. (pl.) 粪便，屎

feedback

n. 反馈

feedlot

n. 饲育场

fence

n. 栅栏，围墙

v. 防护，搪塞

fermentation

n. 发酵

ferrous

a. 铁的，含铁的，亚铁的

fiber
n. 纤维
fiendishly
ad. 极坏地，恶魔似地
fierce
a. 凶猛的，猛烈的，暴躁的
filamentous
a. 丝状的
film
n. 影片，薄膜
v. 生薄膜
filter
n. 过滤器，滤波器
v. 过滤，滤过，渗入
fissure
n. 裂缝，裂沟
v. (使)裂开，(使)分裂
fixation
n. 固定 (法)
flare
n. 闪光，火炬
v. 闪光，闪耀
flea
n. 跳蚤
flexibility
n. 弹性，柔性，柔韧性
flexible
a. 易曲的，能变形的
floc
n. 絮凝物，絮状物
flocculate
n. 絮凝物
v. 絮凝，沉淀
fluctuation
n. 起伏，波动，脉动
fluidize
v. 流态化
fluoride
n. 氟化物
fluorine
n. 氟
fluorosis
n. 氟中毒
flux
n. 通量，流出，熔剂，助熔剂
v. 熔化，流出
foliage
n. 树叶，植物
folly
n. 愚蠢，荒唐事
food chain
食物链
food wet
食物网
forefront
n. 最前部，最前线
forestry
n. 林学，林业
fossil
n. 化石，古物
a. 化石的，陈腐的
foul
a. 污秽的，邪恶的，淤塞的
v. 弄脏，妨害，淤塞，腐烂
ad. 不正当地
n. 犯规，缠绕
fragile
a. 易碎的，脆的
fraught
a. 含有……的，充满……的
freon
n. 氟里昂
frictional
a. 摩擦的
fuel
n. 燃料，木炭

v. 供燃料

fumigation

n. 熏蒸法，烟熏法，熏蒸作用

fundamental

n. 基本原理，基本原则

a. 基本的，重要的，基础的

fungi

n. (p1.) 真菌，霉菌

fungus

n. 真菌，霉菌

funnel

n. 漏斗，烟囱

furnace

n. 火炉，熔炉，加热炉

v. 在炉中烧

fusion

n. 熔融，熔化，聚变，熔合物

G

gallon

n. 加仑

galvanize

v. 通电流于，电镀

garbage

n. 垃圾，废物

gasification

n. 气化 (作用)

gasket

n. 垫圈，垫片

gastrointestinal

a. 肠胃的

gauge

n. 标准度量，计量器

v. 测量，精确计量，估计

gear

n. 齿轮，工具

v. 开动，啮合，连接上

gel

n. 胶滞体，凝胶

v. 形成胶体

genetic

a. 遗传的，遗传学的，起源的

geneticist

n. 遗传学家

geometric

a. 几何学的

geothermal

a. 地温的，地热的

germinate

v. 发芽，萌芽

gill

n. 鳃，散热片，峡谷

v. 用刺网捕

gill-breathing

用鳃呼吸

glaciation

n. 冻结成冰，冰川作用

glacier

n. 冰川

glean

v. 拾落穗，收集

glider

n. 滑翔机，滑行者

glimmer

n. 微光，少许

v. 发微光

glue

n. 胶，胶水

v. 粘合，胶合

gradient

n. 梯度，坡度，斜率

grain

n. 谷类，颗粒，晶粒，纹理

v. (使) 成细粒

grant

n. 授予，授予物，允许
v. 允许，承认，授与
granular
a. 颗粒的，粒状的
grasshopper
n. 蚱蜢，蝗虫
gravel
n. 砂砾，碎石
gray
n. 灰色
a. 灰的，灰色的
v. (使) 变灰色
grazer
n. 食草动物
grease
n. 脂肪，润滑脂，猪油
v. 涂脂于，贿赂
greenhouse
n. 温室
grille
n. 格子，铁格子，格子窗
grind
n. 研磨，磨，摩擦声
v. 磨擦，磨碎
grit
n. 粗砂石
v. 覆以砂砾
grizzly
a. 呈灰色的
n. 灰熊
grouchy
a. 不高兴的，不满的
groundwater
n. 地下水
grumpy
a. 性情乖戾的，脾气坏的

H

habitat
n. 栖息地，聚集处
halogen
n. 卤素
halve
v. 二等分，把……减半
harmony
n. 协调，和睦
hatch
n. 孵化，舱口
v. 孵，孵化，策划
haul
n. 用力拖拉，拖运距离
v. 拖，拖运，改变主意
hawk
n. 鹰，掠夺者
v. 捕捉，咳嗽，散播
haystack
n. 干草堆
hazy
a. 朦胧的，模糊的，烟雾弥漫的
heal
v. 治愈，和解
hectare
n. 公顷
heed
n. 注意，留意
v. 注意，留心
helical
a. 螺旋状的
helmet
n. 钢盔，盔，安全帽
v. 给……戴头盔
hemisphere
n. 半球
henceforth

ad. 今后，自此以后

herbicide

n. 除草剂

herbivore

n. 食草动物

herbivorous

a. 食草的

heritage

n. 遗产，继承权

heron

n. 苍鹭的巢

heterotrophic

a. 异养的

heterotroph

n. 异养生物

hexavalent

a. 六价的

hive

n. 蜂房，蜂巢，闹市

v. 储备，聚居

homogeneous

a. 同种的

honeybee

n. 蜜蜂

hopper

n. 漏斗，贮斗，集尘箱

horror

n. 恐怖，恐惧

horticultural

a. 园艺的

host

n. 主人，宿主，多数

humanity

n. 人类，人性

humus

n. 腐殖质

hurricane

n. 飓风

husbandry

n. 农业，饲养，管理，节俭

hyacinth

n. 风信子，水葫芦，红锆石

hydraulically

ad. 用水 (液) 压的方法

hydrocarbon

n. 碳氢化合物，烃

hydroelectric

a. 水电的

hydroelectricity

n. 水力电气，水力发电

hydrogen

n. 氢

hydrogenation

n. 氢化 (作尾)

hydrogeological

a. 水文地质的

hydrolyze

v. 水解

hydroxide

n. 氢氧化物，羟化物

hydroxyl

n. 氢氧根，羟基

hyena

n. 土狼，贪婪的人

hygroscopic

a. 吸湿的，吸湿性的

hypothetical

a. 假设的，理想的

I

ignition

n. 点火，点燃

illicit

a. 违法的

immense

a. 无边的，非常好的

immune

a. 免疫的，免除的，不受影响的

n. 免疫者

impairment

n. 削弱，减少，损害

impervious

a. 不能透过的，不受影响的

implementation

n. 实施，实现，履行

impurity

n. 杂质，不纯

inadvertent

a. 漫不经心的，非故意的

incidence

n. 发生，影响，倾角，发生率

incident

n. 事件，事变

a. 附带的，入射的，易于发生的

incineration

n. 焚烧，焚化，火葬

incinerator

n. 焚烧炉，焚化炉

incompatible

a. 不协调，不相容，矛盾的

incorporate

a. 合并的

v. 吸收，合并，混合

incredible

a. 难以置信的

increment

n. 增量，增值，盈余

indicative

a. 指标的，表示……的

indicator

n. 指示剂，指示器

indigenous

a. 本土的，国产的，固有的

industrialization

n. 工业化，产业化

ineffectual

a. 无效的，无益的

inert

a. 惰性的，迟钝的

inertia

n. 惯性，惰性

inertial

a. 惯性的

inexorably

ad. 不屈不挠地，坚决地，无情地

infectious

a. 传染性的

infiltration

n. 渗滤，渗透，渗透物

ingredient

n. 成分，因素

inhabit

v. 居住于，栖息

inherently

ad. 内在地，固有地

inhibit

v. 禁止，抑制

injection

n. 注入，注射，注射剂

innovation

n. 改革，创新

inoculate

v. 接种，嫁接

insecticide

n. 杀虫剂

insidious

a. 阴险的，狡猾的

intangible

a. 无形的

integrity

n. 整体性，正直

intention

n. 目的，意图

interbreed

v. (使) 异种交配，(使) 杂种繁殖

intercept

n. 截取，防碍，截距

v. 阻止，截取

interdisciplinary

a. 学科间的

interior

n. 内部，内政

a. 内部的

intermingle

v. 搀杂，混合

intermittent

a. 间歇的

interpret

v. 解释，翻译

interspecific

a. 中间的

intestine

a. 内部的，国内的

n. 肠

intrusion

n. 闯入，侵扰

intuitively

ad. 直觉地

inversion

n. 反转，倒置

invertebrate

a. 无脊椎的，无骨气的

n. 无脊椎动物

iodine

n. 碘，碘酒

iodometric

a. 碘定量的

ion

n. 离子

ionize

v. 离子化，电离

irate

a. 生气的

irradiate

v. 照射，放射，发光

irreparable

a. 不可挽回的

irritability

n. 易怒，过敏性

irritation

n. 激怒，刺激物

isobar

n. 等压线，同重元素

isolate

v. 使隔离

n. 隔离种群

isolation

n. 分离，孤立，隔离

isothermal

a. 等温线的，等温的

n. 等温线

J

jerky

a. 颠簸的，不平稳的

jest

n. 笑话，笑柄

v. 讲笑话，嘲笑

jug

n. 水壶，监牢

v. 炖，关押

jurisdiction

n. 司法权，管辖权

K

kerosene

n. 煤油

kilter

n. 良好状态，顺利，平衡

kinetic energy

动能

knoll

n. 小山，圆丘

L

lace

n. 饰带

v. 扎带子

lag

n. 滞后，落后，防护套

a. 最后的

v. 落后，滞后

lagoon

n. 近海岸的浅水区，咸水湖

laminate

v. 制成薄板

a. 薄板状的

n. 薄片制品

landfill

n. 垃圾填埋法，垃圾

lapse

n. 过失，流逝，失效

v. 犯错，消失，使失效

larvae

n.（昆虫）幼虫，幼体

latitude

n. 纬度

lattice

n. 格子，晶格，点阵

v. 使成格子状

lawsuit

n. 诉讼，诉讼案件

layout

n. 布置，陈列，规划图

leach

v. 沥滤，滤取

n. 过滤器

leachate

n. 渗滤液，沥滤液

lead

n. 铅

leakage

n. 漏，泄漏，渗漏

lean

n. 倾斜，倾斜度

a. 贫乏的

v. 依靠，倾斜，依赖

leftover

n. 残留物，剩饭

a. 残余的

legislative

n. 立法机构

a. 立法的

legitimize

v. 使合理

legume

n. 豆，豆类

leguminous plant

豆科植物

lemon

n. 柠檬

lesion

n. 损害

lethal

a. 致命的

n. 致死因子

lettuce

n. 莴苣，生菜

lichen

n. 青苔，苔藓

liftoff

n. 搬走，卸下，发射

limpet

n. 纠缠者

liquefaction
n. 液化（作用），溶解
liter
n. 公升
literally
ad. 逐字地，不夸张地
lithosphere
n. 岩石圈
litter
n. 垃圾，杂乱，轿，担架
v. 乱丢（垃圾），产仔
livestock
n. 家畜
loader
n. 装填机
lobster
n. 龙虾
lofting
n. 放样，理论模线的绘制
log
n. 航行日志，原木
v. 伐木
logarithm
n. 对数
looping
n. 回路，环
lubrication
n. 润滑油
lush
a. 青葱的，丰富的，豪华的
lyse
v. 溶解

M

macerate
v. 浸渍，浸软
macronutrient
n. 大量营养素
magnesium
n. 镁
majestic
a. 宏伟的
malady
n. 病，疾病，弊病
malaria
n. 疟疾
mallard
n. 绿头鸭
mandatory
a. 命令的，强制性的
n. 代理人
manifest
n. 载货单，运货单
a. 显然的
v. 显示，表明
manipulation
n. 操作，处理
marine
n. 舰队，水兵
a. 海的，海上的，海运的
marmot
n. 土拨鼠
marsh
n. 沼泽，湿地
masonry
n. 砖石建筑
mastic
n. 乳香脂，乳香，胶合铺料
mat
n. 垫子，一丛，编织品
a. 粗糙的，无光泽的
v. 铺垫子
mate
n. 配偶，对手
v. 成配偶，结伴
maturation

n. 化脓，成熟

maxima

n. (pl.)最大量，极大值

meadow

n. 草地，牧场

meld

v. 混合，合并

membrane

n. 膜，隔膜，薄膜

mend

n. 改进，修理

v. 修改，改进

mercaptan

n. 硫醇

merchantable

a. 有销路的

merge

v. 合并，吞没

mesh

n. 网目，筛目，网眼

meteorology

n. 气象学

methane

n. 甲烷，沼气

methanol

n. 甲醇

micronutrient

n. 微量营养素

microscopic

a. 微观的

minor

n. 次要科目

a. 较小的，次要的

miscellaneous

a. 各种各样的，混杂的

mitigate

v. 缓和，减轻

moderate

a. 适度的，中等的

v. 节制，减轻，使缓和

moisture

n. 含水量，湿气

mole

n. 摩尔，克分子

mollusc

n. 软体动物

momentum

n. 动量，要素

monoculture

n. 单一农作物

monoethanolamine

n. 单乙醇胺

monument

n. 纪念碑，石碑

mortality

n. 死亡率

mosquito

n. 蚊子

motherhood

n. 母性，母亲们

mount

v. 安装，设置，增加，安放

mow

n. 草堆

v. 割草

mucus

n. 粘液

muddy

a. 泥泞的，浑浊的

v. 搅混

muff

n. 保温套，失败，笨拙

v. 弄糟，错过

muffler

n. 消声器，围巾

mulch
n. 覆盖物
mule
n. 骡，倔强之人
multilateral
a. 多边的，多国的
multimillion
n. 数百万
multinational
a. 多国的，多民族的
n. 跨国公司
multistage
a. 多级的
municipal
a. 地方性的，市政的
municipality
n. 自治市，市政当局
muskeg
n. (尤指北美北部的) 泥岩沼泽地
mussel
n. 贝壳类
musty
a. 发霉的，过时的
mutualism
n. 互利共生，共栖

N

nasal
a. 鼻的，鼻声的
n. 鼻音，鼻音字
nasopharyngeal
a. 鼻咽的
necessitate
v. 迫使，需要，使成为必要
negotiation
n. 谈判，协商
neon
n. 氖
neutral
n. 中立者，中立国，空挡
a. 中性的，中立的
neutralism
n. 中立，中立主义
neutralization
n. 中和 (作用)，中立状态
niche
n. 壁龛
v. 安顿
nitrate
n. 硝酸盐
nitrification
n. 硝化作用
nitrite
n. 亚硝酸盐
nitrobacteria
n. 硝化菌属
nitrogen
n. 氮
nitrosomonas
n. 亚硝化单胞菌属
nomogram
n. 计算图表
nonferrous
a. 非铁的，无铁的
non-point source
非点源，面源
nullify
v. 作废，取消
nutrient
a. 营养的
n. 营养物

O

objectionable
a. 令人讨厌的
obsession

n. 困扰

octave

n. 八度音阶，八个一组的物品

odorous

a. 有气味的

offensive

a. 令人不快的

offset

n. 抵消，偏移，错位，残留误差

v. 弥补，补偿

offspring

n. 子孙，后代

olfactory

a. 嗅觉的

n. 嗅神经，嗅觉器官

omnivore

n. 杂食动物

onion

n. 洋葱

on-site

现场

onslaught

n. 突击，猛击

opaque

n. 不透明物

a. 不透明的，不传热的

optimism

n. 乐观，乐观主义

ore

n. 矿石

organism

n. 生物体，有机体

orient

n. 东方，东方诸国

a. 东方的

v. 向东方，确定方向

ornamental

n. 装饰物

a. 装饰性的

ornithologist

n. 鸟类学家

orthophosphate

n. 正磷酸盐

oscillation

n. 振动，摆动

osprey

n. 鹗，白色的羽毛

outbreak

n. 爆发，发作

overshoot

v. 过火，越过，过分

n. 越过度

owl

n. 猫头鹰

oxidant

n. 氧化剂

oxidation pond

氧化塘

oxidize

v. 氧化

oyster

n. 牡蛎，沉默者

ozone

n. 臭氧

P

pad

n. 填料，衬垫

v. 填充

palatable

a. 可口的

panacea

n. 灵丹妙药

panel

n. 小组，仪表板，栅栏

parasite

n. 寄生虫
parasitic
a. 寄生的
parasitism
n. 寄生，寄生状态，寄生现象
particulate
n. 微粒
a. 微粒的
partridge
n. 鹧鸪
passive
a. 消极的，被动的
pathogen
n. 病原体
pathogenic
a. 致病的
pea
n. 豌豆
pear
n. 梨子，梨树
pebble
n. 小鹅卵石
peel
n. 果皮，外壳
v. 削……皮
peeper
n. 窥视者，嘀咕的人
pelican
n. 鹈鹕，伽蓝鸟
penetrate
v. 渗透，穿透
per capita
每人
perceptible
a. 能感觉得到的
percolate
v. 滤过，(使) 渗出
n. 渗滤液
peregrine
a. 流浪的，移居的，外国的
n. 寄居外国的人
perforate
v. 刺穿，穿孔于，穿过，开孔
a. 有孔的
peril
n. 危险，险境
v. 冒险，危及
perish
v. 毁灭，毁坏，腐烂，死亡
permafrost
n. 永冻地层
permeability
n. 渗透性，渗透率
peroxide
n. 过氧化物
perpendicular
n. 垂直，垂线
a. 垂直的，直立的
perpetual
a. 永久的，持续的
perspective
n. 透视，透视法
a. 透视的，透视法的
pessimist
n. 悲观者
pest
n. 灾害，鼠疫
pesticide
n. 杀虫剂，农药
pheasant
n. 雉鸡，野鸡
phenol
n. 酚，苯酚
phospholipid
n. 磷脂
phosphorus

n. 磷

photochemical oxidant

光化学氧化剂

photochemistry

n. 光化学

photoexcitation

n. 光致激发，光激

photon

n. 光子，光量子

photosynthesis

n. 光合作用

photovoltage

n. 光电压

physiological

a. 生理学上的，生理学的

piezometric

a. 量压的

pillar

n. 支柱，墩

pine

n. 松树

v. 憔悴，痛苦，怀念

pinto

a. 黑白花斑的，杂色的

pistol

n. 手枪，信号枪

piston

n. 活塞

planktonic

a. 浮游生物

plaque

n. 匾，饰板，斑点

plateau

n. 高地，高原

platen

n. 压纸卷轴，压盘

pliable

a. 易弯的，柔顺的

plume

n. 羽毛，羽流，烟羽

pneumatic

a. 气动的，气体的，气体学的

n. 充气轮胎

point source

点源

polarity

n. 极性，截然相反

pollen

n. 花粉

v. 传授花粉给

pollinate

v. 对……授粉

pollination

n. 授粉

pollutant

n. 污染物

ponder

v. 沉思，考虑

population

n. (生物) 种群，人口

postulate

n. 假定，基本条件

v. 要求，假定

potassium

n. 钾

potent

a. 有力的，有效的

potential energy

势能

prairie

n. 大草原

precarious

a. 不稳定的，危险的

precipitation

n. 析出，沉淀 (作用)

precipitator

n. 沉淀器，除尘器
predator
n. 食肉动物，捕食者
presbycusis
n. 老年性耳聋
prevalence
n. 普遍，流行
preventative
a. 预防的
n. 预防措施
prickly
a. 多刺的
probe
n. 探索，探针
v. 调查，探索
producer
n. 生产者
proliferate
v. 增生，增殖
prolific
a. 多产的，丰富的
prolong
n. 冷凝管
v. 延长，拖延
prominent
a. 卓越的，显著的
propagation
n. 增殖，繁殖，传播
propellant
n. （火箭）推进剂
proprietary
a. 所有者的，专利的
n. 所有人，所有权
protein
n. 蛋白质
a. 蛋白质的
protest
n. 抗议，反对，抗议书
v. 抗议，反对
protoplasm
n. 原生质
protozoa
n. 原生动物
proverbial
a. 谚语的，闻名的
provision
n.（政府提供的）钱和设备，规定
v. 供给……食物及必需品
provoke
v. 激怒，刺激，诱导
pseudomonas
n. 假单胞细菌
psychoacoustic
a. 音质的
puff
n. 一阵喷烟，喘气
v. 喷出，(使) 充气，(使) 膨胀
pulley
n. 滑车
pulmonary
a. 肺部的
pulp
n. 纸浆，果肉
v. 使化成纸浆
pulverize
v. 粉碎，把……磨成粉
pup
n. 小狗，幼畜
v. 生小狗
purification
n. 净化，纯化
putrescibility
n. 腐败性
putrescible
a. 可腐化的，容易腐烂的
pyrolysis

n. 热解，高温分解

Q

quadruple

a. 四倍的

n. 四倍

v. 使成四倍

quaternary

a. 四元的，四价的

quench

v. 淬火，(使) 骤冷，熄灭

quiescent

a. 静止的，寂静的

R

raccoon

n. 浣熊

radioactive

a. 放射性的

radionuclide

n. 放射性核素

radius

n. 半径

ramification

n. 分歧，分支，衍生物

ranch

n. 大牧场

v. 经营牧场

rancher

n. 农场主，农场工人

rarefaction

n. 稀薄，稀疏 (状态)

ratio

n. 比，比率

raven

n. 掠夺

a. 乌黑的

v. 掠夺

reaeration

n. 曝气

realm

n. 领域，领土

rear

n. 后面，背面

a. 后面的，背面的

v. 养育，饲养，栽种

receptor

n. 接受器，受体

recipient

a. 易接受的

n. 接受者，容器

reclamation

n. 开垦，教化

rectangular

a. 矩形的

recycle

v. 使再循环

n. 再循环

redress

n. 赔偿，补救

v. 纠正，赔偿

refinery

n. 精炼厂

refractory

a. 不易处理的，耐火的

refurbish

v. 刷新

rehabilitative

a. 使复原的，复职的

relative humidity

相对湿度

relevant

a. 有关的，相关的

reliance

n. 信任，信赖，信心

remedial

a. 治疗的，补救的

render

v. 汇报，反映，执行，实施

n. 交纳，粉刷，打底

repel

v. 相互排斥，抵制，使厌恶

repercussion

n. 反射，弹回

replant

v. 改种，使移居

replenish

v. 再充满，加强

reptile

n. 爬行动物

a. 爬行的

reradiate

v. 反向辐射，转播

residue

n. 残渣，剩余物

resilient

a. 回弹的，有弹力的

resistance

n. 抵抗，阻抗

resonance

n. 共振，共鸣

resort

n. 常去之处，凭借，手段

respiration

n. 呼吸 (作用)

retention

n. 保持，保留，停滞

retentivity

n. 保持力，记忆力

reverberation

n. 反响，回响，反射

rhizobium

n. 根瘤菌

rhythm

n. 节奏，旋律

ribbon

n. 色带，带状物，缎带

v. 成带状

rigid

a. 坚硬的，刚性的

riparian

a. 河岸的，水滨的

n. 河岸土地所有人

rival

n. 对手，竞争者

a. 竞争的

roast

n. 烤肉，烘烤

a. 烘烤的

v. 烤，焙烧，煅烧，嘲笑

robin

n. 知更鸟

rodent

a. 咬的，侵蚀性的

n. 啮齿动物

rotate

a. 辐状的

v. 回转，旋转，循环

routinely

ad. 日常，乏味，常规

ruinous

a. 破坏性的

runoff

n. 径流 (量)

S

sabin

n. 赛宾 (声吸收单位)

saga

n. 传说

sake

n. 缘故，理由

salable
a. 有销路的，畅销的
salamander
n. 火蜥蜴，耐火的人
salient
a. 显著的，突出的，跳跃的
n. 凸角，突出部分
salinization
n. 盐渍（碱）化
salmon
n. 鲑鱼
sandpiper
n. 鹬（鸟名）
sanitary
a. 卫生的
n. 公共厕所
sanitation
n. （环境）卫生，卫生（设备）
saturation
n. 饱和，饱和度
scavenger
n. 清洁工，净化剂，食腐动物
scent
n. 气味，嗅觉
scrap
n. 碎片，残余物
v. 敲碎，拆毁
a. 废弃的
scrubber
n. 洗涤器，清洁工
scum
n. 浮渣，水垢，泡沫
v. 撇去（浮渣），产生泡沫
seal
n. 印章，封条，标志
v. 封闭，盖章
seam
n. 缝，接缝
v. 裂开，缝合，焊合
seaweed
n. 海藻
sediment
n. 沉积，沉淀物
seed
n. 种子，籽
v. 结实，播种，催……发育
seek
v. 寻求，查找，探索，追求
seep
v. 渗出，漏，渗流
n. 小泉
seepage
n. 渗流，渗漏，渗出物
semipermeable
a. 半透的
septic
a. 腐败的
n. 腐烂物
sequentially
ad. 循序地，从而
settlement
n. 居住区，解决，处理，沉降
settling tank
沉淀池
sewage
n. 下水道（系统），污水
sewer
n. 排水管，阴沟
v. 敷设下水道
shaft
n. 轴，杆，烟囱，塔尖，矿井
v. 利用
shale
n. 页岩
shampoo
n. 洗发，洗发精，洗发香波

v. 洗头，洗发
shear
n. 剪应变，剪切
v. 修剪，割，切断，剪切
shellfish
n. 贝，甲壳类水生物
shield
n. 防护屏，保护者，屏蔽
v. 保护，屏蔽，庇护
shrub
n. 灌木
silica
n. 硅石，二氧化硅
silicon
n. 硅
simultaneously
ad. 同时发生，同时
sink
n. 污水槽，沟渠，洗涤池
v. 下沉，沉没，渗透
sinusoidal
a. 正弦曲线的
skeletal
a. 骨骼的，框架的
skeptic
n. 怀疑论者
skepticism
n. 怀疑论
skimmer
n. 撇沫器
skimming
n. 撇去浮沫，撇清
slat
n. 条板
v. 劈啪劈啪地碰撞，用条板制造
slaughter
n. 残杀，屠杀，屠夫
v. 残杀，屠杀
slime
n. 粘液，烂泥
v. 涂泥，变粘滑
slimy
a. 泥泞的，讨厌的
slot
n. 槽，水沟，缺口，位置
v. 开槽于
slough
v. 脱落，碎落
sludge
n. 软泥，泥状沉淀，污泥
slurry
n. 泥浆
soar
n. 高飞程度
v. 翱翔
sodium
n. 钠
solubility
n. 溶解性，溶度
solvent
n. 溶剂
a. 有溶解力的
songbird
n. 鸣鸟，女歌手
soot
n. 煤烟，烟灰
sophisticate
n. 久经世故的人，老油条
v. 篡改，曲解，弄复杂
sound intensity
声强
soundproof
a. 隔音的
soybean
n. 大豆
span

n. 全长，跨距，范围
v. 跨越，架设，持续
sparrow
n. 麻雀
sparse
a. 稀疏的，稀薄的
spawn
n. (鱼) 卵，产物
v. 产卵，大量生产
specialization
n. 专业化，特殊化
species
n. 种，物种，类
species diversity
物种多样性
spinach
n. 菠菜
spinner
n. 纺纱工人，旋转器
spongy
a. 海绵状的，多孔的
sporadic
a. 偶尔发生的，分散的
spore
n. 孢子
v. 长孢子
spray
n. 喷雾器，水沫，浪花
v. 喷淋，喷，喷雾，喷射
spruce
n. 云衫
v. (使) 干净，打扮整齐
squirrel
n. 松鼠，松鼠毛皮
v. 贮藏
stabilize
v. 稳定，使安定
stack
n. 烟囱，堆
stagnate
v. (使) 淤塞，(使) 停滞
standpoint
n. 立场，观点
steer
v. 引导，驾驶，控制
n. 驾驶指示，劝告
stiffness
n. 坚硬性，稳定性
stimuli
n. 刺激，刺激物，促进因素，
stipulate
v. 规定，保证
stirrup
n. 钢筋箍，U 形卡
stockpile
n. 储蓄，积蓄
v. 贮存，储蓄
stoker
n. 加煤机，司炉
stopper
n. 塞子，阻凝剂
stranglehold
n. 抑制，束缚
stratigraphy
n. 地层学
stratosphere
n. 平流层，同温层
stringent
a. 迫切的，严厉的
strip
n. 长条，带，脱衣舞
v. 脱衣，剥夺，拆卸
subdivision
n. 分部，细分
submerge
v. 使浸水，淹没

submicron

n. 亚微米

substantial

n. 重要部分，要领

a. 物质的，真正的，相当大的

substitute

n. 代理，代理人，代用品

v. 代替，取代

subtle

a. 敏锐的，稀薄的，灵巧的

subtract

v. 减去，扣掉，减少

subtraction

n. 减少

succinct

a. 简洁的，紧缩的

suffice

v. 足够，有能力，满足

suffocate

v. 使窒息，噎住，受阻

sulfate

n. 硫酸盐

sulfide

n. 硫化物

sulfur

n. 硫，硫磺

superimpose

v. 叠加，添加，附加

supplementary

a. 辅助的，补充的，追加的

surplus

n. 剩余，过剩，盈余

a. 过剩的，剩余的.

susceptible

a. 易受影响的，容许...的

suspended solids

悬浮固体

suspicious

a. 可疑的，多疑的，怀疑的

sustainable

a. 可持续的

swamp gas

沼气

sway

n. 摇摆，支配，影响力

v. 摇动，支配，影响

swirl

n. 旋涡，旋转器

v. 使成旋涡，打漩，盘绕

switchgear

n. 开关设备

synthetic

n. 合成物质

a. 人造的，合成的，综合的，接合的

T

tag

n. 标签，附属物，标记，附加语

v. 尾随，连接，紧紧跟随

tamper

v. 干预，损害，夯实

n. 夯具，填塞工具

tan

n. 棕褐色

a. 棕褐色的

v. 晒成棕褐色

tangentially

ad. 成切线

tannic

a. 丹宁酸的，鞣酸的

taper

n. 圆锥形，尖锥形

a. 渐尖的，锥形的

v. 弄尖，递减

teaspoonful

n. 一茶匙的容量

tenuous
  a. 稀薄的，脆弱的，纤细的
terminology
  n. 术语，专用名词
tern
  n. 三个一组，燕鸥
terrain
  n. 地带，地区，地形，领域，范围
terrarium
  n. 小植物栽培盒
terrestrial
  n. 地球上的人
  a. 地球的，陆生的，陆地的，人间的
tertiary
  n. 第三位的，第三产业的
  a. 第三纪
tertiary treatment
  三级处理
textile
  n. 纺织品，纺织业
  a. 纺织的
thermodynamic
  a. 热力学的
thicken
  v. 变浓，加强，增稠
threshold
  n. 开端，阈，阈值
tint
  n. 色彩，色调
  v. 染色于
tiny
  a. 很少的，微小的
titration
  n. 滴定，滴定法
toad
  n. 蟾蜍，癞蛤蟆，讨厌的家伙
tobacco
  n. 烟草，香烟
topography
  n. 地形学，地势
topsoil
  n. 表层土
tornado
  n. 旋风，龙卷风，大雷雨
tortuous
  a. 扭曲的，弯曲的，曲折的
total solids
  总固体
tout
  v. 推销，侦察
  n. 兜售者
toxic
  a. 有毒的，中毒的
toxicity
  n. 毒性
toxin
  n. 毒素
tracheobronchial
  a. 支气管的，气管的
tract
  n. 广阔地面，地方，小册子
trailer
  n. 拖车，追踪者
trajectory
  n. 轨道，轨线
transition
  n. 转变，转换，变迁，过渡
transparent
  a. 透明的，清晰的
transpiration
  n. 蒸发，蒸腾，排出
trap
  v. 截留，捕获，收集
  n. 收集器，格栅，闸门
trench
  n. 渠，沟渠，战壕

v. 掘沟，挖战壕

trickling filter

滴滤池

trigger

n. 触发器，扳机

v. 触发，引发，引起，发射

trihedral

a. 有三面的

trim

n. 整齐，情形，修剪

a. 整齐的

v. 使整齐，整理，调整，修剪

trioxide

n. 三氧化物

tritium

n. 氚

triumph

n. 凯旋，胜利，欢欣

v. 得胜，成功

trivalent

a. 三价的

trophic level

营养级

troposphere

n. 对流层

trout

n. 鲑鱼

tub

n. 桶，浴盆

v. 装入桶，洗澡

tundra

n. 苔原，冻原

turbid

a. 浑浊的，污浊的

turbidity

n. 浊度

turbulence

n. 湍流，紊流，混乱

twofold

a. 两倍的，双重的

ad. 两倍地

tympanic

a. 鼓室的，鼓膜的

typhoid

a. 伤寒的

n. 伤寒

U

ultraviolet

a. 紫外线的

undercoat

n. 底漆，下层绒毛，内涂层

undue

a. 不适当的，过度的

unleash

v. 解开，解放，发动

unsaturated

a. 不饱和的，未饱和的

unsound

a. 不稳固的，不健全的，腐烂的

upgrade

n. 上坡，升级，上升

ad. 往上

v. 使升级，提升，改良品种

upshot

n. 结果，结局，(论证的) 要点

urbanization

n. 都市化

urine

n. 小便

V

vagrant

n. 流浪汉，无赖

a. 漂泊的，流浪的

vane

n. 轮叶，风向标

vanish

v. 消失

vapor

n. 汽，蒸汽

v. 蒸发，使蒸发

vector

n. 矢量，载体，向量

v. 给……导航

vegetation

n. 植物

vengeance

n. 复仇，报复

vent

n. 排气口，通风孔，发泄

v. 发泄，排出

ventilate

v. 使通风，

ventilation

n. 通风，换气

venturi

n. 文氏管，文丘里管

vermin

n. 害虫，歹徒，寄生虫

versus

prep. 对……

vessel

n. 船，容器

via

prep. 经过，经由，通过

vibration

n. 振动， 颤动

vibratory

a. 振动的， 震动的，振动性的

vicinity

n. 附近，邻近

vigil

n. 警戒，监视，守夜

vigor

n. 精力，活力

vinyl

n. 乙烯基

violate

v. 违反，亵渎，侵犯，妨碍

viral

a. 病毒的，病毒引起的

virtually

ad. 事实上

virus

n. 病毒

viscoelastic

a. 粘弹性的

viscous

a. 粘的，粘性的

visibility

n. 可见度，能见度

void

n. 空隙，孔率，真空

a. 空的，无效的，无用的

volatile

a. 挥发性的，不稳定的，爆炸性的

n. 挥发物

volatile solids

挥发性固体

volatilization

n. 挥发 (作用)

volatilize

v. 挥发，使挥发

volcanic

a. 火山的，猛烈的

n. 火山岩

vole

n. 田鼠，大满惯

v. 全胜

voltage

n. 电压，伏特数

vomit

v. 呕吐，吐出

n. 呕吐，呕吐物

vulnerability

n. 易受伤，易损性

vulture

n. 秃鹫，贪婪的人

vying

a. 竞争的

W

wafer

n. 圆片，薄酥饼

wanton

a. 恶意的，胡乱的，放纵的

v. 闲荡，挥霍

warrant

n. 授权，许可证，证明，批准

v. 授权给，保证，担保，批准

wastage

n. 废物，浪费，浪费量

waste disposal

废物处理

water supply

供水，给水

waterborne

a. 水传播的，水运的

watercourse

n. 水道，河道，流，水路

waterlogging

n. 涝，水浸

waterway

n. 航道，水路，水道

wear

v. 磨损，消耗，耐久，呈现着

n. 损耗量，衣服

wetland

n. 湿地，沼泽地

wholesale

n. 批发

a. 批发的，大规模的

ad. 大规模

v. 批发

windshield

n. 挡风玻璃，风遮蔽

wiper

n. 雨刷，滑动片，弧刷

woodchuck

n. 土拨鼠

wrought

a. 精致的，锻制的

Z

zeal

n. 热心，热衷，热诚

zinc

n. 锌

v. 镀锌于

zooplankton

n. 浮游动物

References

[1] U.S. Agency for International Development. Climate Change in Vietnam: Assessment of Issues and Options for USAID Funding, Final Report[R]. 2011.

[2] United Nations Development Programme Climate Change and Human Development Report Office, Occasional paper[R]. 2007.

[3] Asian Institute of Technology/United Nations Environment Programme, Regional Resource Centre for Asia and the Pacific. Scoping Assessment on Climate Change Adaptation in Vietnam Summary–Regional Climate Change Acknowledgement Platform for Asia, Bangkok, Thailand[R].2010.

[4] Intergovernmental Panel on Climate Change. Climate Change 2007: Synthesis Report – An Assessment of the Intergovrnmental Panel on Climate Change, IPCC Plenary XXVII[R]. Valencia, Spain, 12-17, 2007.